A wonderful and haunting book: so rich in detail that the mammals of Britain's past seem brought to life again, and yet so unsparing in its portrayal of the brute facts of extinction that readers will ache for all that has been lost.

Tom Holland, author and historian

As elegies go, *The Missing Lynx* is an awful lot of fun. ... ...g is clear and unobtrusively witty ... ...rry Practchett-esque footnotes.

...es

Rewilding is certainly romantic ... ...ll needs thinking through, and Bar... ...r of the idea, can help.

BBC Wildlife

Fresh and assured ... An often moving tribute to lost marvels.

Nature

The story of Britain's Ice Age bestiary, told with bittersweet humour, and a clarion call to us all to step up and fight future extinctions.

Tori Herridge

Barnett's writing is clear and accessible, and often amusing.

BBC Countryfile

With his fast-paced and amusing tales of some of the most awe-inspiring species lost within geologically recent times, *The Missing Lynx* brings Britain's Ice Age back to life.

Professor Beth Shapiro, author of *How to Clone a Mammoth*

A fascinating account of the large herbivores and predators that have disappeared from Britain since humans reached our islands. This should be essential reading for those who advocate rewilding.

Professor Richard Fortey, palaeontologist and author

The first chapter of @DeepFriedDNA's new "The Missing Lynx" is one of the most powerful, emotionally arresting openings I've read in a book in a long, long time.

The Well-read Naturalist

Read this book by palaeontologist Ross Barnett – it's quirky, informative, superbly well-written and crammed with an avalanche of facts (and utterly fascinating digressions) which bring a tantalising world of recently vanished ice age creatures to life. And did I mention the footnotes? This book has THE most entertaining footnotes of any book I've ever read!

Ian Roberts, author

It is a fascinating publication by a specialist in analysing and interpreting ancient DNA. Packed full of scientific data and detail of fossil discoveries ... it is written with great humour, even cunningly weaving in snatches from literature and film.

Professor Keith Somerville, *talkinghumanities*

A NOTE ON THE AUTHOR

Ross Barnett is a palaeontologist with a PhD in Zoology from the University of Oxford. He specialises in seeking, analysing and interpreting ancient DNA, but his area of expertise is the genetics and phylogeny of cats, especially the extinct sabretooths. Barnett's research has led to many remarkable findings in recent years and has involved investigating escaped lynx in Edwardian Devon, rubbishing claims that the yeti is an ice-age polar bear and seeking the ancestral home of the enigmatic Orkney vole. In 2018, he received the Palaeontological Association's Gertrude Elles Award for Public Engagement. Barnett currently lives in the Highlands of Scotland with his wife and two daughters.

@DeepFriedDNA

THE MISSING LYNX

The Past and Future of Britain's Lost Mammals

Ross Barnett

BLOOMSBURY WILDLIFE
LONDON · OXFORD · NEW YORK · NEW DELHI · SYDNEY

BLOOMSBURY WILDLIFE
Bloomsbury Publishing Plc
50 Bedford Square, London, WC1B 3DP, UK
29 Earlsfort Terrace, Dublin 2, Ireland

BLOOMSBURY, BLOOMSBURY WILDLIFE and the Diana logo are
trademarks of Bloomsbury Publishing Plc

First published in Great Britain 2019. Paperback edition 2020.

To find out more about our authors and books visit www.bloomsbury.com
and sign up for our newsletters

Dedicated to my hopes for the future:
Alice Iona and Naomi Catriona

Contents

Prologue

> The crust of our earth is a great cemetery, where the rocks are tombstones on which the buried dead have written their own epitaphs.

> Louis Agassiz

We are living through the sixth mass extinction.

Stretch your arms out either side of your body. Go on. Nobody's watching. Now, imagine that the distance from your left fingertip to your right fingertip is the history of all complex life on earth, starting with the first multicellular animal* about 580 million years ago.

Using this bodily chronometer, the first mass extinction takes place at your left elbow. Here, at the end of the Ordovician period 444 million years ago, nearly nine-tenths of all marine species die out. The disappearance of many forms of graptolite, preserved as tiny comb-shaped fossils found worldwide, give testimony in the rocks. Life continues.

The second mass extinction takes place at your left shoulder. Here, at the end of the Devonian period 376 million years ago, three-quarters of all species disappear. The coiled ammonite and the segmented trilobite barely scrape through. Life continues.

---

* If your stretched-arm timeline were to start with the first cell, multicellular animals would only begin to appear between your right elbow and wrist.

The third mass extinction occurs at your right shoulder: this was the big one around 250 million years ago, known as 'the Great Dying'. The end of the Permian period nearly sees the extinction of all life on earth. Only four out of every hundred species survive. The few trilobite groups that had lived through the Devonian extinction curl up and disappear completely. No one really understands what happened to cause it, but a massive increase in greenhouse gases, especially methane, created by new forms of bacteria, is one suspect. Life continues.

The fourth mass extinction takes place at your right elbow. Here, at the junction between the Triassic and Jurassic periods 200 million years ago, only half of all species disappear. Conodonts, diverse members of a lamprey-like family of boneless fish, disappear completely, leaving nothing but their enigmatic teeth behind. Life continues.

The fifth mass extinction, at your right wrist, is the end of the dinosaurs' reign 66 million years ago. This transition between the Cretaceous and Palaeogene periods witnesses the demise of the dinosaurs,[*] flying pterosaurs and marine ichthyosaurs and plesiosaurs. The impact of an asteroid at the Chicxulub crater site in Mexico is strongly implicated in the process. Death from space. Millions of tonnes of vaporised bedrock entering the atmosphere and making the sun as black as sackcloth. Global cooling, acid rain, mass extinction. Life continues.

[*] Except birds, which are actually a surviving lineage of dinosaur. If you ever doubt this truism, go and watch some cassowaries.

The sixth extinction starts at about where the nail on your right middle fingertip began growing this morning. Those few micrometres of keratin stand for the last 50,000 years of death on earth. A time during which modern humans spread through the world and we see the extinction of mammoths and moa, dodos and diprotodonts, pampatheres and passenger pigeons, toxodonts and thylacines. Our infiltration of every corner of the planet and the disappearance of so many species is not a coincidence.

We are causing the sixth extinction.

CHAPTER ONE

The Past

At midnight in the museum hall
The fossils gathered for a ball
There were no drums or saxophones,
But just the clatter of their bones,
A rolling, rattling, carefree circus
Of mammoth polkas and mazurkas.
Pterodactyls and brontosauruses
Sang ghostly prehistoric choruses.
Amid the mastodontic wassail
I caught the eye of one small fossil.
'Cheer up sad world', he said, and winked –
'It's kind of fun to be extinct.'

Ogden Nash, 'Fossils'

The story of life on earth is the story of extinction. It is recognition that all is flux. *Memento mori*: remember that you die! It should humble us. Whispering insistently into our ear that Ozymandias's fate is our fate too, extinction tells humanity that sooner or later we either change and become something else, or fall into the abyss.

For us, the story of extinction begins with the realisation that mammoths were not just elephants washed north by Noah's flood. It took a huge conceptual leap to abandon the Bible as a guide and trust the primacy of geology and rocks. Nevertheless, the discoveries of frozen mammoths and giant sloths demanded it. Species were not divinely created but had appeared and disappeared at various and remote periods of the past. Some of the greatest minds of the eighteenth and nineteenth centuries obsessed over relics of bones and the monsters that made them. Many pondered a simple question that we are only just beginning to answer: What happened?

The geological epoch we live in today is called the Holocene. The previous epoch is known as the Pleistocene, and the two together make up the Quaternary period. In the great calendar of geological time, epochs are like minutes and periods are like hours that count the days of succeeding eras. You might be more familiar with the Pleistocene under its alternative and more visceral name of the Ice Age.

Starting about 2.5 million years ago, the earth lurched from a lush, tropical paradise to a planet capped by extensive ice to the north and south. During the Pleistocene, the waxing and waning of the ice occurred in distinct pulses. These pulses are seen in records of global temperature preserved in ice bubbles in Greenland and Antarctica. Cylinders of laminated ice painstakingly drilled out of the ground by scientists give a year-by-year account of the earth's climate. The ice cores show how the pendulum has swung many times between warm and frigid. The Ice Age was colder overall than today but alternated between long stretches of cold and brief warm respites. The graph of temperature over time looks uncannily like a heartbeat as recorded on an electrocardiogram.

Periods of freezing conditions when summer temperatures were many degrees cooler than they are today are known as 'glacials'. At the height of the

Pleistocene glacials, which lasted for 100,000 years, glaciers many miles thick swaddled North America from Fairbanks to New York. The Great Lakes and Hudson Bay are the sparse remains of millions of tonnes of ice, and the great gouges they took from the earth. In Eurasia, similar forces made glaciers coalesce into ice sheets completely covering Scotland, Ireland, Wales, northern England, Scandinavia and the Barents Sea. Mountains of ice carved the fjords, planed the mountains and scraped the sea floor.

Then, regular as clockwork, the climate switched again. During the warm interglacials, global temperatures were similar to today and the ice retreated, waiting patiently to rise again. The Holocene, where we are now, is really nothing but a particularly stable interglacial period. We are still in the Ice Age, albeit one that has been on pause for the last 10 millennia. It's pure hubris on the part of humanity to give our little interglacial its own named epoch. Really, nothing quantifiable climate-wise separates the Holocene from previous interglacials. The details of why the earth has swung between two extremes for the last couple of million years were first tackled by a seemingly unlikely pair of individuals working decades apart.

A self-educated* Scotsman named James Croll was the first to realise that variations in Earth's orbit around the

* A huge understatement. Croll had little formal education but managed to get a job as a janitor at a Glasgow museum. From there he corresponded with Darwin, Lyell and other Victorian scientists to develop his ideas and finally got an academic position at the Geological Survey of Scotland. Croll's story reads like a nineteenth-century *Good Will Hunting*.

Sun, caused by the gravitational pull of other planets, would affect the amount of solar radiation received in different seasons. The Earth's orbit around the Sun is not perfectly regular but alternates from almost circular to nearly oval. This is known as eccentricity of orbit and follows a cycle that repeats every 100,000 years. Eccentricity causes changes in the amount of sunlight reaching Earth in winter, which in turn affects snow cover; if snow persists into the summer, it reflects more light, cooling the planet and causing a feedback cycle, meaning more and more snow can accumulate, initiating glacial conditions.

Later, a very smart Serbian mathematician called Milutin Milankovitch improved on Croll's work to include further variables and provide a better fit between theory and data. No small matter. It took him 20 years to complete all the calculations, by hand. Milankovitch's contributions related to how the Earth itself wobbles on its axis. Two different wobbles happen. Over 41,000 years, the tilt of the Earth's axis (obliquity) varies by around three degrees, which changes the distribution of sunlight over land and ocean. Working with a periodicity of 21,000 years, the axis also moves around a point like a spinning top about to tip over. This spin is known as the precession of the equinoxes and affects which season occurs when the Earth is closest to the Sun. Together, eccentricity, obliquity and precession explain why our world has regularly experienced glacial conditions, and just as regularly snapped back into interglacial conditions. Over more than two million years, the species of life on earth have passed through innumerable phases of cold and warm. The Holocene is just the most recent.

Of course we are all familiar with a cartoon version of what the Ice Age would have looked like: walls of ice, thick carpets of snow, biting winds. Whether you immediately think of Hoth, Castle Black or *Ice Age: Dawn of the Dinosaurs*, they are all equally wrong. Yes, on top of the glacial ice sheets you could have found those conditions, and for a few miles in front of them too, but otherwise full glacial periods might have surprised you with their abundance of plant and animal life. One of the most counter-intuitive findings is that actually much of the northern latitudes not covered in ice were highly productive grasslands: a habitat known as the 'mammoth steppe'. Thanks to increased aridity, reduced cloud cover, heavy manuring by mammoths, rhinos and other giant herbivores, and grazing pressure keeping trees in check, a vast area from France to Alaska was one continuous belt of highly nutritious forage. Also, thanks to the millions of tonnes of water locked into glaciers and ice sheets, global sea levels were many metres lower than they are today. The area between Siberia and Alaska, known as Beringia, was dry grassland. The North Sea did not exist. Australia, Papua New Guinea and Tasmania were connected. Borneo, Java and Sumatra were joined to Malaysia. Animals and people could walk to places that are now islands.

For Britain, being connected to mainland Europe made it a peninsula of France, not an island. With no English Channel, megafauna could easily walk from Paris to London. This is certainly the route that both Neanderthals and modern humans took to Britain. Where the North Sea is today would have been vibrant grassland 50,000 years ago, dotted with herds of woolly mammoths. Cave lions would have fought with cave hyaenas over the carcasses of cave bears. Giant Irish elk would have lorded it over valleys full of woolly rhinos and steppe bison. Neanderthals might have skulked along the margins. Today, bones of all these species are dredged up by fisherfolk trawling the seabed between Rotterdam and Grimsby. Neanderthal skulls, mammoth tusks, sabretooth jaws and flint tools have all been caught side by side with cod and haddock. The bones are stained night-black from the decaying peat they have sat under for millennia. Often they are covered in barnacles or cold-water coral. The fisherfolk donate many of them to museums. To augment their incomes, some of the commoner trawls are sold on eBay or other online platforms to supplement the harsh conditions of the fish markets. I have in front of me right now a beautiful example of the metatarsal (foot bone) of a cave lion, plucked from the seabed 15 years ago in a fruitless quest to extract DNA.

We know from dating the bones that turn up in the nets that many of them are very old, from the Pleistocene through to the early Holocene – the sea only slowly claimed the land after the ice began to melt. The fishing spot known as Dogger Bank, a shallow depression approximately halfway between Edinburgh and Copenhagen, was the last part to succumb to the rising waves. When dry, 'Doggerland', as it has been called, may have been a kind of northern Atlantis, an island occupied by seasonal people up until the Mesolithic (9,000 years ago). By this time most of the megafauna (animals heavier than 44 kilograms or 100 pounds in old money) and all the megamammals heavier than 1,000 kilograms or 2,200 pounds had disappeared from Europe.

The crunch point for megafauna is the transition from the Pleistocene to the Holocene. Geological mavens set it at 11,700 years ago. What is different about the Holocene is not the climate but the biota. Fifty thousand years ago, every continent had its equivalent of the African Serengeti. Everywhere was teeming with wild animals. In comparison, most of the modern world is like a desert. On the grasslands of Pleistocene California, herds of American bison mingled with enormous Columbian mammoths and elephant-like mastodons. Harington's horses were stalked by sabretooths. Shaggy giant sloths, skittish yesterday's camels and wary Shasta ground sloths kept a nervous eye on prides of American lions and dire wolves on the lookout for their next meal. In Pleistocene New South Wales, marsupial lions harried 2.5-metre (8-foot) kangaroos and 2-tonne wombats. Rainbow snakes as thick as a person's waist, giant ripper lizards and massive horned tortoises left them to it while a surfeit of other giants lumbered on. In Pleistocene Patagonia, sabretooth cats, giant jaguars and enormous short-faced bears brought the remains of cow-sized sloths, extinct giant llamas and deep-nosed horses back to their dens for easy consumption. Animals belonging to families that have disappeared so completely they have no surviving relatives to compare them to lolloped on three-toed hooves amongst southern beech trees – entire boughs of the tree of life gone forever.

The world of today seems terribly meagre in comparison. So much is gone. In the Pleistocene you have intact ecosystems everywhere. In the Holocene the continents are stripped bare. We see the nature of what we have lost in the bones of giants, dragons and monsters[*] dug out of the ground and found in caves and fissures. Painstakingly pieced together following the rules of comparative anatomy, megafaunal skeletons grace the great halls of museums all over the world. With a bit of imagination it is possible to see the living animal. Drape the bones in muscle and skin and fur and breathe life in. Think about how it ate: Did it graze the grass or browse the treetops? Think about how it found a mate. Go crazy imagining vivid colour schemes or elaborate courtship rituals lost to time. With the Pleistocene megafauna you can rest assured that somebody knew the truth. All these extinct behemoths were seen by people like you and me. In special cases our ancestors left a record of what the actual animal looked like in life. Think of the painted caves and ivory carvings of mammoths, rhinos and the like that have come down to us from the Palaeolithic. In even rarer cases we have the mummies of mammoths and their kin, unearthed from the permafrost of distant Siberia.

I am here as an advocate for the extinct Pleistocene megafauna. Dinosaurs have their fans, and that's fine if you like that sort of thing. Personally, dinosaurs can never have the appeal to me that the megafauna have. Despite what the creationists insist, no person ever saw a *Tyrannosaurus*

---

[*] There is strong circumstantial evidence that the myth of the Cyclops Polyphemus, encountered by Odysseus, is based on discoveries of fossil elephant skulls – the large nasal socket looks like a giant central eye. This theory is explored at length in Adrienne Mayor's *The First Fossil Hunters*.

rex or a *Triceratops horridus.** On the other hand, every single extinct Pleistocene species interacted with humans. People saw them, people knew them and people ate them. Your great (times a thousand or so) grandparents would have known what colour a sabretooth cat was. They would have known the best way to cook mammoth steak. They would have known how to avoid cave bears waking from hibernation. The megafauna were part of our world and we were part of theirs. 'Were' being the operative word. The megafauna are mostly gone, and the megamammals disappeared from everywhere except Africa and South Asia. Why? There are three main arguments that have been used and dozens of fringe beliefs.†

The first is that they could have disappeared due to disease. Some new virus or bacteria that could infect mammoths and mastodons, sloths and sabretooths, could have spread like the plague through susceptible communities. Jumping from species to species and with high fatality, disease could have decimated the giants. There are hints of similar plagues in the modern world. One example is

---

* The binomial (Latin name) system consists of two parts: the first, the genus, denotes that all members of that group have a recent common ancestor, like the genus *Panthera* for lions, leopard, jaguars, tigers and snow leopards. The second part gives the species name (in this case *leo, pardus, onca, tigris* and *uncia,* respectively). Species can be subdivided further into subspecies that represent different pockets of diversity and adaptations to different habitats. The level above genus is family – in our *Panthera* case, that would be Felidae (aka 'cats'). See the appendix for a fuller explanation and a list of all the species mentioned in this book.

† A recent vogue has been for extraterrestrial impact of a comet or meteorite having a role in the extinctions (yawn). The data for this is not convincing to any serious person.

chytridiomycosis, a fungal infection that is wiping out
frogs, toads, newts and every kind of amphibian it
touches. It seems to be a new disease but one that has
spread successfully to every continent that has amphibian
life. Similarly, white-nose syndrome sweeps through bat
colonies, irritating resting bats out of winter torpor and
making them starve to death. Currently, these diseases
are wreaking devastation on wild populations but are
not causing mass extinctions. So far.

The problem with this scenario is that there are very
few potential diseases that could infect mammoths and
rhinos and bison and horses and all the other extinct
megafauna. Such a wide range of potential hosts needs
an infectious agent that would be devastating indeed.
Tuberculosis has popped up in the bones of some ancient
bison and mastodons, but this just emphasises the main
problem. Most diseases do not have an effect on the
bones, and are therefore invisible in the fossil record.
They are issues of tissues. Where we have complete
mummies, no obvious signs of infection are found either.
Although, only those living within a decade or so of the
extinction event would be expected to have any sign of
the causative agent. There is also the added complication
that congenital diseases are at high prevalence in small,
inbred populations, as would be expected to occur in
any species approaching extinction from any means.
And why could we not still find the fatal disease
occasionally in the giant mammals of today?

The second argument looks to environmental
conditions. Changes in climate are the leitmotif of the
Pleistocene. Climate change seems an obvious culprit for
the extinction of the megafauna. There are, however,

some complications. As previously noted, the switch between the Pleistocene and the Holocene was not special. The temperature changes were no more extreme than those at 100,000, 200,000 or 300,000 years before. The same species (mammoth, rhino, short-faced bear, sabretooth, giant sloth, *Diprotodon*) sailed through previous climate challenges, only happening to go extinct near the end of the Ice Age at different times in different continents. In the same way, all the large species we have now – bison, red deer, kangaroo, tapir – did not go extinct. Whatever happened was selective. It took the stag-moose but not the moose, the American cheetah but not the African cheetah, the woolly rhino but not the Sumatran rhino, the cave bear but not the brown bear.

Why? Climate does not cause selective extinction like this, especially when many of the extinct species occupied completely different niches. Woolly rhinos on the Beringian steppe, giant sloths on the Patagonian pampas, cave bears in the alpine meadows. It doesn't make any sense. Many researchers are very keen on the idea of climate change at the Pleistocene–Holocene transition causing the wave of extinctions. I have yet to see a model that explains the selectivity, or why similar changes earlier in the Pleistocene did not unleash the same havoc.

Two extinct groups in particular cause massive problems for the idea that climate alone could have caused extinction. Woolly mammoths lived along the belt of mammoth steppe from Spain to America before dying out just about everywhere by 10,000 years ago. They survived for another 7,000 years on Wrangel Island, just 130 kilometres (80 miles) off the Russian coast. Woolly mammoths overlapped with the flourishing of Egyptian civilisation,[*] were around after the invention of writing and are literally younger than Moses. What possible change in climate could eliminate mammoths on the mainland, yet leave them in perfect harmony just 80 miles away? None. The disappearance of the Wrangel mammoths coincides closely with the first archaeological evidence of humans on Wrangel.

Giant sloths in dozens of genera, from bear- to elephant-sized, disappeared from the Americas 11,000 years ago. They survived on the islands of Cuba and Hispaniola for another 6,000 years. Large sloths (*Megalocnus rodens*, a 200-kilogram or 440-pound beastie that was the largest animal in the Caribbean) and medium-sized sloths (*Neocnus comes*) only vanish around 5,000 years ago. Again, how could climatic factors totally wipe out sloths on the mainland but leave island relatives only 160 kilometres (100 miles) away unscathed? They couldn't. But archaeology shows that the Caribbean was settled at the same time that the sloths disappear.

There is some evidence that the cyclical changes in climate stressed out populations of megafauna, causing them to migrate to places where conditions were more favourable, or squeezing them into refugia where they could shelter and ride out the swings in temperature. But there is nothing – no model, no explanation – for why climate change would cause extinctions of large, slow-reproducing species worldwide.

---

[*] There is a deliciously tantalising image of a hairy elephant in the tomb of Rekhmire, a minor official during the reign of Thutmose III. Some have argued that it could be a Wrangel mammoth; others that it could be from one of the Mediterranean islands that may have had elephants until the Middle Holocene.

And the third argument is us. The way I see it, there is only one difference at the end of the Pleistocene that could explain what happened: the spread of modern humans with a suite of behavioural and technological adaptations for which the megafauna was totally unprepared. For many reasons, the push and pull between advocates of climate change and human responsibility for extinction have been particularly acrimonious. Plenty of mud has been slung in the usually staid pages of major scientific journals.

For me, this is the only explanation that makes sense. And it isn't a new idea. It's been around since Pleistocene extinction was recognised.* While floating in the aether for more than a century, the idea was given its biggest injection of energy from 1967 onwards by the redoubtable Paul S. Martin,[†] who coined the term 'overkill' as a snappy buzzword to encapsulate his ideas.

Let's walk through the main strands of data. First is the issue of timing. The Quaternary extinctions did not happen simultaneously but were spaced out, with different regions collapsing at different times. Our grasp

---

* Alfred Russel Wallace, co-discoverer of evolution by natural selection, was a proponent. He wrote in his 1911 publication *A World of Life*: 'I am convinced that the [...] rapidity of the extinction of so many large Mammalia is actually due to man's agency, *acting in co-operation with those general causes* [emphasis in the original] which at the culmination of each geological era has led to the extinction of the larger, the most specialised, or the most strangely modified forms.'

[†] Martin was a fascinating character. It is doubtful whether 'overkill' would have anywhere near the power it has in the market of ideas without his forceful contributions. His knowledge of Pleistocene mammals was unparalleled.

of the chronology now is much better than when Paul Martin was first working in the 1960s. Barring a few background extinctions, the first noticeable wave of concentrated loss happens in Australia. The ancestors of Aboriginal peoples arrived here 47,000 to 50,000 years ago. Hugging the coast on their multigenerational meander from Africa, they must have made some perilous open ocean journeys from island to island before reaching Oz. When they arrived, they found a wonderland of marsupial giants. Enormous wombats, ridiculous-sized kangaroos, flightless birds with rugby-ball-sized eggs ... By 45,000 years ago all were gone. Evidence for direct predation (those eggs made plenty of omelettes, leaving burnt eggshells behind) and human landscape modification (fire drives and controlled burnings) are still left in the landscape. Tasmania, colonised slightly later (at 43,000 years ago) by people who walked over a dry Bass Strait, held onto its megafauna until 41,000 years ago.

Northern Eurasia had been home to the Neanderthals for 200,000 years before modern humans invaded their territory. Nonetheless, there do not seem to be any extinctions associated with their arrival. Anatomically modern humans burst into Eurasia around about 40,000 years ago and a staggered series of extinctions is evident. Cave hyaenas and cave bears disappear at the same time as the Neanderthals (who were, all things considered, megafauna as well). Later on, cave lions, woolly mammoths, woolly rhinos and Irish elk go missing around 14,000 years ago.

In North America, humans traipsed over a dry Bering Strait, the steppe grassland of Beringia, and entered North

America perhaps as early as 35,000 years ago. They couldn't get any further than Alaska or Yukon, though, because miles of ice covered the whole of Canada in the form of the Cordilleran and Laurentide glacial ice sheets. Between 16,000 and 15,000 years ago, hugging the Pacific coast, they ventured south into new lands and new horizons. They multiplied and spread, developing new cultures (including the fearsome Clovis weaponry: flint AR-15s hafted to throwing spears), and moving ever southwards. By 11,500 years ago all the mammoths, giant sloths, horses, glyptodonts and camels were gone.

In South America, humans arrived less than a millennium after breaking south of the Canadian ice sheets. By 11,000 years ago all the giant elephant-like gomphotheres, bear- to giraffe-sized giant sloths, armoured tank-like glyptodonts, weird notoungulates and litopterns were dead. South America lost 48 genera of megafauna, more than any other continent.

For all the continents, a freaky coincidence is evident. People turn up and megafauna disappear.

Secondly, there is evidence of direct hunting. Skeletons of mammoths have been found with flint points stuck between the ribs and buried deep in vertebrae. Mastodons have been found with spearpoints made from the bones of other mastodons protruding out of their ribs.* A gomphothere from Venezuela had a

---

* This is the famous Manis mastodon. It was found in Washington State and at 13,800 years old is one of the first signs of humans in continental America. The Manis mastodon has a rib that has been pierced by a sharp bony projectile, also made from mastodon bone. It's clear evidence of the hunting prowess of the first immigrants to enter the Americas.

stone projectile wedged into its pubic bone, and another from Mexico had its bones piled in heaps with stone spearpoints scattered throughout. Horse and yesterday's camel bones from Wally's Beach in southern Canada were found covered in cut marks and mixed with lithic debris. At the very tip of South America, in the appropriately named province of Last Hope (Última Esperanza) in Chile, a giant cave preserved the remains of ground sloth bones, dung and skin from the Pleistocene. The pelt, complete with hair and the unique bony dermal ossicles that made it as impenetrable as chain mail, was clearly skinned from a butchered animal by Pleistocene humans. [*]

There are other sites with other extinct species, but it's important to recognise that it's miraculous there are any at all. One of the predictions of Martin's model is that evidence of Pleistocene hunting by modern humans should be sparse. It happened over a short period of time, over 10,000 years ago. Initial preservation is unlikely. Continued survival of sites is unlikely. All these unlikelihoods should add up to not much evidence at all. And yet, there is.

---

[*] The skin has had an incredible history. Discovered in the late nineteenth century by locals, it spent time hanging on a tree as a curio. The great Swedish explorer Otto Nordenskjöld cut off a portion and took it back to Uppsala. Word spread and many others took their share, including Francisco Moreno, a famous Argentinian pioneer of palaeontology. Now, the skin (or skins) are to be found in museum stores in London, Berlin, Leiden and elsewhere. A small piece makes a cameo appearance in Bruce Chatwin's seminal *In Patagonia* as the coveted 'skin of a brontosaurus' that his relative Charley Milward sent to his grandmother. To Bruce's eternal regret, his mother threw it out after his grandmother died.

Unfortunately, the materials with which humans made the vast majority of their hunting weapons – wood and bone – do not survive very well in the archaeological record. There are, however, exceptions. The Middle Pleistocene Clacton spear from Essex and the Schöningen lances from Hanover show that for hundreds of thousands of years, even pre-humans had the ability to create pointy, fire-hardened sticks capable of killing megafauna. Flint, being stone, is much better represented in the archaeological record, and this informs much of what we think about what people could do. In North America, a few thousand years after people pushed south of Canada, they developed a new culture – if the difference in their material goods is any indication. Called Clovis, after the New Mexico site where it was first recognised, it is dominated by enormous, complex stone points. Typically around 15 centimetres (6 inches) long, and looking for all the world like a bodyboard or longboard made out of stone, they are exquisite objects. Symmetrical and aesthetically pleasing, they were obviously designed for hafting to wooden spears. And they are deadly. It's no idle speculation; archaeologists have done the legwork. One in particular, George Frison, made replica Clovis points and tested them on culled African elephants in the wilds of Zimbabwe. Using an atlatl,* and from as far away as 20 metres (65 feet), Frison was easily able to use Clovis technology to accurately penetrate the skin, bone

---

* An atlatl is a spear-thrower that is just a piece of hooked wood, allowing the spear to be thrown with greater force – the ancient equivalent of those plastic things you see dog owners use to throw balls.

and organs of dead elephants. Clovis points, which let's remember have been found associated with the bones of megafauna, are perfectly capable of felling megamammals. These points ain't for hunting rabbits. There are thousands of them too, found all over North America. It's like coming across the aftermath of a battle with dead bodies all around, riddled with bullets and guns strewn everywhere, then asking: 'How did all these people get killed?'

Thirdly, there is the evidence of islands. Islands can be continents writ small; unique ecosystems with evolutionary novelties that are found nowhere else. The dodo of Mauritius lived for aeons on its island home with giant tortoises, flightless parrots and other wonders. Never having met predators like humans, they had no defences, and written records from the seventeenth century describe with glee how easily Dutch sailors could capture and kill the animals.

New Zealand was the same. Before discovery by seafaring Polynesians in the thirteenth century, there were no land mammals except bats. New Zealand was a birdly paradise. The giant moa lived here as buffaloes and antelopes live in Africa, grazing and browsing. Moa were relatives of the ostrich, emu and kiwi – entirely flightless[*] and very big. The largest moa, the South Island giant moa, could easily reach 2.7 metres (9 feet) in height and 250 kilograms (550 pounds) in weight.[†] Within a couple

---

[*] Moa are the only bird family to lack any wing bones – they've lost them completely! The lack of buffalo wings is entirely made up for by the size of the drumsticks, though.

[†] Just the females. Amazingly, the difference in sexes was enormous, with females twice as big as the comparatively puny males.

of centuries of human arrival, all 11 species were gone, even the smallest turkey-sized species, along with the giant New Zealand eagle that preyed on them and several other large, flightless birds. No written records were left behind, but the story is told by the enormous middens of burnt and broken moa bones found at many sites throughout New Zealand. As only a few hundred years have passed since moa extinction, their presence is expected; their remains tell an eloquent tale of over-exploitation for meat and eggs, along with the impact of deforestation and feral animals (the ancestors of the Māori also brought rats and dogs with them). Māori language has many pithy aphorisms, replete with references to species not seen by any living person for hundreds of years. The idea of what living moa were like survives only in the language. One saying still in use is 'ka ngaro I te ngaro o te moa': 'lost as the moa is lost'!

Madagascar, the large island off Africa's east coast, was colonised earlier than New Zealand. There is controversy over when exactly, but people were definitely on Madagascar by 350 BC. Today, Madagascar is famous for its lemurs, and they are a mere reflection of what used to be there: gorilla-sized giant lemurs, floppy-eared pygmy hippos, giant tortoises and two enormous flightless birds. Today, beaches in Madagascar may contain millions of fragments of eggshell from the giant elephant bird, scattered like broken crockery on the sands that were its home. Extinction in Madagascar seems to have taken longer than in New Zealand. Some species may even have survived until the sixteenth or seventeenth centuries, when Madagascar became a French colony. Étienne de Flacourt, a French governor, wrote evocative stories about a bird called the vouronpatra that was always hiding: 'So that the people of these places may not take it, it seeks the most lonely places.' Similar tales of cow-like animals with the face of a man (giant lemurs?) are known. There is even ethnological evidence of twentieth-century survival of hippos in the interior of the island (unlikely but perhaps a record of an accurate folklore of this species).

The modern period has only seen the extinction of one megamammal. Steller's sea cow was an innocuous, innocent, gentle grazer of the kelp meadows of the North Pacific. Part of the same family as dugongs and manatees, this enormous beast could be 9 metres (30 feet) long and weigh up to 10 tonnes. In Pleistocene times it had ranged from Japan to California, wherever kelp was to be found. In recent times its entire home was a few patches of shoreline around the Commander Islands of the Bering Strait, east of Kamchatka. It would have been safe in this remote outpost, far away from people. It should have been safe from us.

It wasn't. In 1741, the great German naturalist Georg Steller (who lent his name to half a northern biota) discovered the sea cow – the only biologist to witness the species alive. Steller found employ with the Russian government on a ship captained by Vitus Bering[*] on a voyage of discovery to Russia's eastern limit. Nominally

[*] While Georg was busy naming animals – Steller's sea cows, Steller's sea lions, Steller's jays, Steller's eiders and Steller's sea eagle – Bering was having geographical features named after him: Bering Island (where he is buried) and Bering Strait (hence Beringia).

a voyage to map and explore the fringes of Russia's empire and whether it joined the Americas, in reality it was also an attempt to find lucrative fur trapping and whaling grounds. In this they succeeded but at a cost. The entire ship was wrecked on uninhabited Bering Island, and Bering himself and many of the crew died soon after from scurvy. The sea cow literally saved the survivors' lives. The animals were utterly naive to the danger that humans posed; the starving sailors were able to row up to them with a billhook and slaughter them where they lay. The tonnes of meat each sea cow provided kept Steller and his shipmates alive long enough to survive the winter and build a new boat from their wreckage to take them back to civilisation. Steller, making observations even while starving and shipwrecked, described the sea cows' meat as beef-like and delicious, the blubber exquisite and buttery, even raw. He even milked dead nursing females for their 'rich and sweet milk'. The animals had no defences from knives and hooks, and could be taken at leisure. Steller wrote that the peaceful cows would pathetically try to aid their injured comrades, thrashing boats with their tails and trying to pull hooks and spears from the bodies of their friends. Family groups would graze together on the shore, the calves in the middle of the group. Steller noted that only one calf was born per year, with pregnancy lasting perhaps 18 months. Monogamous pairs would stay by their wounded partners, even while the sailors were busy flensing* its dying remains.

---

* What does it say about our species that we need a specific verb to describe the flaying of a marine mammal?

Unfortunately for the sea cows, Steller and the survivors returned to mainland Russia with stories of innumerable blue foxes and sea otters just waiting to be exploited. A throng of trappers made straight for the Commander Islands, where they hunted the foxes and otters for their fur and killed sea cows for food. It took just 27 years from first discovery for the sea cow to be utterly wiped from the earth. By 1768 none were left.

The histories of Mauritius, New Zealand, Madagascar and especially the Steller's sea cow are the Rosetta Stone to deciphering what went on in the Pleistocene extinctions. Unique, specialist species with little to no experience of hunting by humans and narrow ecological niches were exterminated. Large megafauna and megamammals that had slow reproductive rates were decimated.

Once your mind has grasped the idea that the Pleistocene extinctions were caused by humanity, it makes you appreciate them in a different way. If it's a combination of climate and humans then it's clear where the fault lies. Only one of these causes has agency, foresight, thought.

What this realisation should show is that the word 'natural' is meaningless in the modern world. Outside of Africa and South Asia, there are no natural habitats. Everything has been modified, changed, degraded. In American conservation circles, it used to be a given that the aim was to restore the national parks and wilderness to a pre-European balance. There hasn't been balance for over 10,000 years! The recognition that there is no Eden, no original state of grace, takes some getting used to. Even now, we have an inability to recognise what belongs where in the natural world. Every day there is attrition and loss. A phenomenon known as 'shifting baseline syndrome' means that we can never appreciate what we are losing, even as we are losing it. Shifting baseline happens because human lifespans are so short in comparison to the timescale of the effects that humans have. A child thinks that what they grow up with is the way things should be. Any extinctions that happen during that lifetime are deemed an unnatural tragedy. That person's children will grow up never knowing the extinct species and its absence will be the new normal. We see this happening today with the twentieth-century eradications of lions in Iran, lynxes in Spain and bears in California. You could say that all modern conservation works with the shifting baseline. We are pouring millions of pounds into saving the tiger in India, Nepal and Russia, but already the knowledge that less than a century ago they lived on Bali, Java and around the Caspian Sea in Iran, Turkey and Turkmenistan is more or less ignored. And it will continue – the tiger is practically extinct in China and Sumatra. The species is suffering death by a thousand cuts, losing populations all the time, losing a recognised subspecies every couple of decades. Each generation of conservationists must fight against a recalibration of the norm.

So, what is normal? What is natural? You could argue that a natural Britain is one that has lynxes and wolves, beavers and bears. Or you could slide further back in time and see a Britain with bison and reindeer, moose and wild cattle. Visit a Britain without people and imagine mammoths and rhinos, hyaenas and lions, sabretooths and giant deer. I don't have an answer. All I know is that the Britain and Europe of today is mostly bare of wildlife.

There are serious discussions over whether the world has been permanently damaged by humans enough to consider a new geological era: the Anthropocene. Scientists are hashing out where to define the start of a new epoch. Working backwards, the spikes in global values of rare atomic isotopes from hydrogen bomb tests in the 1950s and 1960s have been mooted. Levels of plastic waste, which will define humanity for geological ages, could act as a marker of the twentieth century. The appearance of high concentrations of elemental aluminium in the geological column is another telltale sign. Concrete slathered over every surface is as clear a boundary as volcanic tephra. Increases in carbon particles from the start of the Industrial Revolution settle on the ground as a discrete layer. There is a telltale decrease in global carbon dioxide around AD 1610 because of the genocide of Native Americans after contact with Europeans, leaving their crops lying uncultivated. All these critical moments have been put forward as start points for the Anthropocene. I think that a better solution would be to just recognise the human causes of the Pleistocene extinctions and synonymise the Holocene with the Anthropocene.

Evolution by natural selection is the ultimate Rube Goldberg machine. Solutions are cobbled together out of whatever is at hand. A stray mutation here, a chance mutation there; building slowly on top of each other, crafting a niche out of thin air. Amazingly, ecosystems are formed spontaneously out of everyone jostling for position. A dash of predation, a soupçon of symbiosis, a pinch of parasitism.* Maybe a sprinkle of jerks, just to keep things interesting. Where intact webs still exist, interactions between and amongst species fuel the incredible diversity and vibrancy of life. Remove any one link in the chain and things start to break down. It may not be immediately obvious; it may take some time, but every extinction matters, and is noted. The great chain of being cannot tolerate too many missing links.

This book is my attempt to speak for the extinct using Britain as a microcosm of what has been lost – to rage against the shifting of the baseline. The species described in each chapter differ in terms of when they were exterminated, and how certain we are of our role in their loss. Nevertheless, I defy anyone to place a boundary between what we know was human-caused and what was not. There is no shifting baseline here. Extinction is forever.

---

* No species is an island, entire of itself. Whenever a species goes extinct, its specialist parasites go with it. Steller recounts finding enormous tapeworms in the dissected entrails of the sea cow. The bark-like skin on their back was encrusted with a kind of amphipod whale-louse that lived in weeping sores and was picked off by opportunistic seagulls. Who mourns for them?

CHAPTER TWO

Cave Hyaena

Extinct in Britain after
c. 46,000 BC; extinct
globally c. 29,000 BC

High on a rock, which o'er the raging flood
Reared its bleak crag, *The Last Hyaena* stood.
Beneath his paws a kindred skull was seen;
And he, with commons short, looked grim and lean.
[...]
But e'er it rose to mix him with the rest.
Thus did he growl aloud his last bequest:-
'*My skull to William Buckland I bequeath.*'
He moaned – and ocean's wave he sank beneath.

Anonymous student of William Buckland,
'The Last English Hyaena'

The cave hyaena (*Crocuta crocuta spelaea*) is one of the most common Ice Age fossils found in Britain. As strange as it may sound to modern ears, hyaenas are native to Britain and have only been absent for a comparatively short time. They were here when people first arrived half a million years ago, and they only disappeared in the Late Pleistocene, about 31,000 years ago. It's no understatement to say that cave hyaenas have had a unique and crucial role in our understanding of the past.

To understand extinction you have to know that fossils come from animals that were alive in the prehistoric past. This conceptual leap is only really possible if you understand the geological context of the fossils. And to really understand geology you have to be taught it. The first-ever lectureship in geology was held by a man who spent a lot of time thinking about British hyaenas: the Reverend William Buckland.

Buckland was a fantastic character. Merely calling him eccentric doesn't really give him the credit he's due. He was a pioneer of eccentricity, forging new directions of (mostly) harmless weirdness. Like many of the great scientists of the nineteenth century, he trained in theology before taking up an academic position at Oxford, where he was able to indulge in his many hobbies. His most obvious quirk was zoophagy: a desire to eat as many different species as possible. Few have been able to match the exotic fare laid on at the Buckland household: mouse on toast,* roast puppy, crocodile, giraffe, rhino, mole,† bluebottle‡ and a suite of dishes from every branch of the tree of life. He even passed this and other obsessions on to his son Francis,§ who became a noted zoophagist himself. All this subjective gastronomic data could have had no better place than in Buckland's inquisitive and experimental mind. Famously, on visiting the holy site of a martyr's inerasable bloodstain, he bounded onto all fours, lapped at the liquid and declared: 'I can tell you what it is: it is bat's urine!'

---

* Apparently quite tasty.
† Apparently pretty disgusting.
‡ The worst.
§ Francis was famous for having exotic pets while a student at Oxford. He had a bear called Tiglath-Pileser, or Tig for short, which he would take to parties or lectures in a custom-made gown and cap. Tig was eventually rusticated and ended up in London Zoo after raiding a sweet shop.

Buckland's association with cave hyaenas and the effect they had on him (and geology) can be tied to one hugely important site. In Yorkshire, near the town of Kirkbymoorside, lies Kirkdale Cave. Buckland, as part of his geological interests, visited Kirkdale in December 1821 after quarrying had exposed the entrance in summer of that year. Inside he discovered an undisturbed fossil wonderland caked in mud; bones of hippo, rhino, deer, mammoth, bear and other species were found. Buckland's careful observation and experimentation[*] showed that unlike the ammonites, ichthyosaurs and other Mesozoic fossils he had worked with, these bones were still organic, and therefore much more recent. By far the most numerous remains were those of hyaenas.[†] There were also some strange things about the fossils discovered in Kirkdale. All the bones had a nibbled appearance; hardly any were complete. Some were shiny and polished on the surface. Balls of weird, white, lumpy material that Buckland called *album graecum* were all over the place. What did it all mean?

Nobody was better prepared to puzzle it out than Buckland. Using the same experimental mind that encouraged him to drink bat piss from the floor of a cathedral, he had some ideas for how to experimentally test his thinking. Back in Oxford he called on the owner of a travelling menagerie who had a spotted hyaena. In a

---

[*] Buckland dissolved some of the bones in acid to remove the calcium phosphate and was shocked to find that the gelatin (i.e. collagen protein) was left behind. This is exactly the first stage used in modern laboratories to extract collagen from bones for radiocarbon dating or stable isotope analysis.

[†] At least 300 canine teeth.

feat of genius, this deeply weird Anglican minister[*] invented the field of experimental palaeontology by feeding pieces of cow to a hyaena and greedily picking up what was left afterwards. He saw that the massive, crunching jaws explained the lack of complete bones in Kirkdale. Spotted hyaenas are bone crackers and relish the chance to split bones with their massive teeth to get at the marrow. They incessantly chew at the edges of bones to get every last scrap of protein, even swallowing the smaller shards. After passing through their guts, the end results are round, bone-dense turds.

Kirkdale was a hyaena den! Generations of cave hyaenas had passed in and out,[†] depositing their faeces all over the place (the *album graecum*), and leaving the bones of their meals strewn about for Buckland to find aeons later. The use of Kirkdale by living animals was as certain as anything could be.

For the deeply religious Buckland, this now posed a conundrum. Prior to his work it was a reasonable and defensible position to say that the action of Noah's flood could explain all the weird bones found in the caves and fissures of Europe. The deluge had washed the elephants and rhinos and hyaenas and hippos and lions and sundry other creatures from their proper home in the tropics, into the nooks and crannies of the north. Kirkdale was proof of something different. Kirkdale showed that these exotics had made their home in England. Had lived and

---

[*] Buckland, for all his eccentricities, seems to have been very well liked (except by Darwin, who called him 'coarse' and a 'buffoon').
[†] This also explained the polishing of the bones. Innumerable cave hyaenas walking over the top had buffed them like the cobbles on a street.

died in Yorkshire. They belonged here. Yes, perhaps Noah's flood had washed in the mud that now covered their bones, but the living animals had stalked England's green and pleasant land. What had their life been like?

Cave hyaenas are significantly bigger than modern spotted hyaenas. Straight comparisons of fossil and modern bones show this unequivocally. It's a recurring theme in our study of extinction. Alfred Russel Wallace, co-discoverer of evolution by natural selection, put it best when he said in his *The Geographical Distribution of Animals*: 'We live in a zoologically impoverished world, from which all the hugest, and fiercest, and strangest forms have recently disappeared.' British cave hyaenas are known from at least the Middle Pleistocene, with bones and fossil excrement found in the coastal sites of East Anglia. Apart from Kirkdale, hyaenas are abundant in the caves of Kents Cavern, Creswell Crags, Paviland, Cefn, Wookey* Hole, Ffynnon Beuno and Bleadon.

How cave hyaenas made it to Britain is an interesting and complicated story. Today, spotted hyaenas are only found in Africa, south of the Sahara desert, which acts as a natural geographic barrier to their spread. It's important to note that the Sahara has not always been there and has repeatedly switched between dry desert and lush grassland, allowing many species, including hyaenas (and

---

* Yes, this is where Chewbacca gets his species name. In George Lucas's feature film debut, *THX 1138*, one of the actors improvised a line about 'running over a wookie' as an in-joke using his friend's surname. George Lucas kept it in mind for his next film, a small independent production called *Star Wars*. The surname is a locational name originating from Wookey. In the film, Chewie's voice was made from recordings of bears, lions and badgers but alas, not hyaenas.

Map 1: *UK and Irish sites where cave hyaena bones have been found, along with calibrated radiocarbon dates where they are known.*

our own species), to travel north. Fossil cave hyaenas are found in Europe, Asia and as far east as the Russian Pacific coast. Cave hyaenas were super-successful predators with a near-global distribution.

Luckily, and thanks to their abundance, researchers have looked at the genetics of British cave hyaenas to work out what was going on. What the genetics show is a complicated pattern of expansion and contraction that is only hinted at in the fossils. Using genetics to work out the history of population movement is known as phylogeography. By comparing DNA sequences from different individuals and different populations, you can see which are related. By comparing this to their geographic origin, it becomes possible to find connections.

The genetics are clear in showing that the spotted hyaena, *Crocuta crocuta*, has its origin in Africa (just like humans). Today there are two genetically separate groups of spotted hyaena: one group is spread across Africa in a band from Senegal in the west to as far east as Somalia, and the other can be found in the southern parts from Tanzania down to the Cape.

When you add in the genetics of fossil hyaenas, the picture evolves. Fossils from the Russian Far East represent one massively distinct branch of the family tree. This could potentially signify a distinct species that has been called *Crocuta ultima*. Over in the West, the European fossils are part of two different groups. Some European fossils belong to their own unique and distinct genetic lineage; others belong to a group very closely related to the modern equatorial African hyaenas.

What all this tells us is that Africa has acted as a kind of hyaena pump, occasionally disgorging spotted hyaenas into the rest of the Old World. This is first thought to have occurred perhaps as far back as the Pliocene (3.5 million years ago) with the exodus of the ancestors of *Crocuta ultima*. Much later, during the Middle Pleistocene (1.5 million years ago), the ancestors of the unique European lineage made their way north. The last hyaena eruption (300,000 years ago) had individuals closely related to the modern African hyaenas move northward. This story closely mirrors our own, with an African genesis and a complicated diaspora.

The genetics of British hyaenas were investigated using some teeth from Creswell Crags* in Derbyshire. They belong to the group that were a close relation to modern African hyaenas. Since there are fossils of British Middle Pleistocene cave hyaenas, and palaeontologists have enough of them to identify gaps in their occurrence, it seems pretty likely that these early pioneers went extinct, and new colonisers reappeared later and then went extinct again. Think of it like our own species. Neanderthals came to Britain, lived happily for a while, and then modern humans came along. Same thing with cave hyaenas. What were they doing when they lived and died in British caves? What were the living animals like? We are lucky in that we can use studies of the modern African species as a guide.

The Hyaenidae are a compact lineage. Today there are only four species left in the family: the spotted hyaena, the striped hyaena, the brown hyaena and the aardwolf.†

Spotted hyaenas get a bad press. From Aristotle to Hemingway, they have been libelled as idiots, scavengers, cowards, gluttons, grave robbers and tricksters. Their distinctive vocalisations just seem to rub *Homo sapiens*

---

* Open to the public, Creswell Crags is an amazing site that displays some of the hyaenas, hippos, lions, horses and other fossils found in a great little museum, as well as the skeleton of a baby cave hyaena called 'Eric'.

† Not just good for scrabble, the aardwolf is the weirdest of the Hyaenidae. It lives entirely on termites, has stumpy teeth and dens underground.

the wrong way (well, no one likes to be laughed at ...). As always, the reality is more complicated, and spotted hyaenas are nothing like the image painted of them.

Spotted hyaenas live in groups of related females. These clans are incredibly hierarchical, with multiple generations of fraternally and maternally related females vying for rank. Even the lowest female usually outranks the highest male. The boys born to the clan don't stay long and are generally pushed out after a couple of years. They wander on to different clans as emigrants, while immigrant males come in to make up the numbers. All this jostling for position, and mental computation of who is related to whom, who needs to be placated and who needs to be scorned, makes for big brains. Spotted hyaenas are not stupid. They're up there with primates and corvids in terms of intelligence. You may have seen videos of problem-solving crows accessing food by opening locked boxes. Well, hyaenas can do the same. Experiments leaving meat in bolt-locked crates on the Maasai Mara in Kenya showed that, like a scarier version of *Jurassic Park*'s raptors, they could work out how to slide the bolt and open the door to get at the meat – without first seeing how it was done. Not stupid. Cave hyaenas would have been the same.

Perhaps the most persistent lie about hyaenas is that they are scavengers. Unlike the noble lion or the majestic cheetah, hyaenas don't hunt but just skulk around other kills waiting to take advantage. Wrong again. With the weight of prejudice against them (thanks, Aristotle!*) it's

---

* 'It is exceedingly fond of putrefied flesh, and will burrow in a graveyard to gratify this propensity' is just one of the things he says.

only comparatively recently that we've come to realise how amazing a predator the spotted hyaena is. Before scientific field studies were put in place, most observations of natural wild behaviour were reports from big-game hunters. They never saw hyaenas hunting.* They often saw lions with game and hyaenas on the periphery – more evidence of their sneaky perfidy! In fact, lions are far more likely to scavenge hyaena kills than hyaenas are to scavenge lion kills. Those peripheral hyaenas were probably waiting to get their meat back. In Africa, spotted hyaenas are the most successful and abundant large predator, and take a wide range of prey, from rodents to wildebeest and zebras. They can work together in their clans to cooperatively take down much larger animals.

In Pleistocene Britain, there would have been plenty of red deer, boar, horse, bison, and other herbivores to hunt. We even have some experimental evidence of cave hyaenas' prey choices. Apart from the gnawed and splintered bones left in places like Kirkdale, modern science has had a go at assessing their diet. Using cave hyaena coprolites† (fossil faeces), as in Buckland's *album graecum*, it has been possible to extract DNA that identifies red deer as a prime constituent of the diet of Pleistocene hyaenas found in France. Cave hyaena

---

* They mostly hunt at night. Mostly.
† Reverend Buckland invented the term 'coprolite' (from the Latin *copro*- meaning filth, and the Greek -*lithos* meaning stone). He published extensively on coprolites and even had a dinner table made that was inset with polished coprolites. This was not usually discussed until after his guests had eaten upon it. I guess it shows you can polish a turd, as long as it's old enough.

coprolites are incredibly resilient and preserve extremely well, even down to their DNA. It appears they are almost rock-solid when 'fresh'. That's why they often turn up in the fossil record. Detailed studies of British cave hyaena coprolites have shown them to be, on average, about a third larger than those of modern spotted hyaenas. They've even been scooped up from the bottom of the North Sea.

Modern African spotted hyaenas have been known to occasionally hunt humans, particularly children sleeping outside at night. The same might have happened with the cave hyaena. At the French site of Arcy-sur-Cure, there is the fossil maxilla of a Neanderthal that has the classic hyaena gnawing seen at Kirkdale, and wherever cave hyaena dens are uncovered. Perhaps our loathing of hyaenas has its origins in the dark and dangerous Ice Age nights.

Having said that, it is entirely possible that the unfortunate Neanderthal who ended up in pieces in a French cave could have been dead before the hyaena ate them. The slur of grave robber is often thrown at spotted hyaenas. It's true that in some African cultures, corpses are deliberately left out in hyaena territory for disposal – to my mind, a fantastic and ecologically sound method of recycling. However, other cultures are less environmentally conscious and go to great lengths to prevent consumption of cadavers, including using stones, bricks and cement to proof the gravesite. The reality is that no carnivore is going to spurn a free meal, but generally hyaenas of all species avoid humans whenever they possibly can. They have far more to fear from us than we do from them.

Now, we turn to the most surprising label given to hyaenas since classical times: that they are hermaphrodites. Unless you've spent a lot of time observing spotted hyaenas, this one takes a little background to understand. The first naturalists to write seriously about spotted hyaenas were confused that the groups seemed to only be made up of males. Only with detailed dissection does the secret emerge. Female spotted hyaenas have a *pseudopenis* that is as large as the penis in the male. The pseudopenis consists of a greatly enlarged clitoris that mimics the position, size and erectile capabilities of the male. In female African spotted hyaenas, dominance is determined by standing nose to tail while the individuals lift their rear legs to allow inspection of the pseudopenis. The subordinate will then lick the dominant. It is a structure unique to spotted hyaenas.

Unlike in most mammals, the urethra passes through the hyaena clitoris as a urogenital canal with which they urinate, copulate and give birth. The mimicry goes further. Female spotted hyaenas have no external vulva and the labia have actually fused to form a *pseudoscrotum* with the exact appearance of the same in the male. The illusion is so complete that even veteran field biologists can find it extremely difficult to sex hyaenas in the wild. With one caveat. I said that spotted hyaenas give birth through their clitoris. As you can undoubtedly imagine, this is a difficult and laborious process. The pseudopenis can be 15 centimetres (6 inches) long or more, and about 2.5 centimetres (1 inch) wide. But the head of a newborn cub has a diameter of around 6 centimetres (2.3 inches). The maths doesn't add up, and it's probably best if I cite verbatim from a 1994 paper by Frank and Glickman on an observation of birth in a captive animal: 'In the birth process, the meatus abruptly tore along the fraenum of the clitoris, allowing passage of the neonate.' The easiest way to identify female spotted hyaenas on the plains, if they have previously given birth, is by looking for the telltale marks of jagged scars on the pseudopenis, where the clitoris has previously torn from trying to pass a baby through a tube that is manifestly too small. And if you think that's traumatic, hyaenas often give birth to twins. These feisty little fuzzballs come into the world with their eyes open and their teeth ready. Massive levels of testosterone make them highly aggressive, and there is some data to suggest that sibling aggression is high and that siblicide could account for the deaths of a quarter of all twin births.

Cave hyaenas and spotted hyaenas are similar enough (apart from size) in their skeletons that they are grouped in the same species. We can use spotted hyaenas as a good proxy for the cave hyaena's life appearance and biology. How similar would they have been? I asked my friend and colleague Professor Lars Werdelin, a world expert in the evolution and fossil history of the Hyaenidae, what he thought. Lars is a laconic and knowledgeable authority on all things hyaena. He cautioned that we shouldn't assume that the complicated social structure of modern spotted hyaenas would be found in fossil cave hyaenas. Since, in the Pleistocene, cave hyaenas would have had much more biomass available for scavenging, they may not have had to hunt so often and this would have knock-on effects for sociality. Perhaps they would have lived in smaller clans, with smaller territories to defend, since the pickings were richer. If anything, I think this makes them more terrifying.

For a super-smart, adaptable, populous predator, it is extremely odd that this species goes extinct, as far as we can tell, before any of the other megafauna. Our best estimate is that cave hyaenas disappear approximately 31,000 years ago. This is just before the height of the Ice Age, in a period known as the Late Glacial Maximum (or LGM, if you're into that whole brevity thing). During the Late Glacial Maximum, things got as cold as they possibly could: glaciers were at their zenith, sea levels at their nadir. We know that cave hyaenas disappeared early thanks to work that systematically looked at their bones and dated as many as possible.

Radiocarbon dating really is crucial to work on the Pleistocene. When it first came about in the 1950s, it was a total game-changer. The theory goes like this: the Sun, being a massive nuclear fusion reactor at the centre of the solar system, is constantly spitting out subatomic particles in the direction of Earth. In Earth's protective atmosphere, sometimes a neutron (a nuclear particle with no charge and an atomic mass of one) will collide with an atomic nucleus of nitrogen (which normally has seven protons and seven neutrons in its nucleus and an atomic weight of 14). When this happens, the nitrogen disgorges a proton so that it now has six protons and eight neutrons. Since we've given names to the elements based on their properties, and this depends mostly on how many protons they have, that nitrogen-14 is transformed into carbon-14, an unstable isotope of carbon. Since it has too many neutrons to be happy, the carbon-14 occasionally emits a beta particle (an electron), which changes that extra neutron into a proton to balance everything out again and turn it back into nitrogen-14. We call emitting a high-energy beta particle by the much scarier name of 'radioactivity'.

In the atmosphere, this process is more or less continual, ensuring a constant ratio of carbon-14 to normal carbon (carbon-12). Of course, all that atmospheric carbon will take part in chemistry – getting turned into carbon dioxide, absorbed by plants, built into their tissues, etc. As long as the plant (or animal that eats it) is alive, the ratio of carbon-14 inside it mirrors the ratio found in the atmosphere. However (this is the cool part), once an organism dies the carbon-14 isn't being topped up with new atmospheric carbon, and because it is unstable and radioactive, the carbon-14 that was in the tissues slowly gets turned back into nitrogen-14. This radioactive decay occurs randomly but in a mathematically certain way, which is called the half-life. Luckily, the half-life of carbon-14 is about 5,500 years. That means that 5,500 years after death, half of your carbon-14 will have turned back into nitrogen-14; after 11,000 years, three-quarters of your carbon-14 will have decayed; after 16,500 years, seven-eighths will have gone, and so on. Carbon-14 can thus act as an atomic egg timer, ticking away time since death. All you have to do is measure the amount of carbon-14 in your sample, compare it to the atmospheric ratio and work out how much has decayed, and that will give you the approximate time since death.

Absolutely any organically preserved remains will work. This mostly means bone but can also mean hair, mummified tissue, wood, paper, etc. However, it doesn't work on actual fossils, since, you know, they're made of rock not bone.* And it doesn't work on organic remains older than about 50,000 years because by that stage you are trying to measure a ratio of 0.1 per cent of the original amount of carbon-14.

The guy that worked all this out, Willard Libby, revolutionised the fields of archaeology, palaeontology, Egyptology and other -ologies. Before radiocarbon dating, there was no way to know how old something was except by comparison. You could dig up a fossil and work out that it was older than the things on top of it and younger than the things below it, but you couldn't give it an *absolute* age. Libby changed all this. He got the Nobel Prize for his work. His name lives on in the 'Libby half-life'.

Annoyingly, when he was doing the work to accurately determine the half-life of carbon-14, he got it slightly wrong. He worked it out as 5,568 ± 30 years, when actually it was more like 5,730 ± 40 years. This was only corrected years after people had been using radiocarbon dating to date their artefacts. In a beautifully human response to the discrepancy, scientists collectively said 'ah, sod it' and continued using the old (and still wrong) Libby half-life in their work. For this (and other) reasons, you tend to get dates described as both 'radiocarbon years' (using the Libby half-life) and 'calendar years' (where the correction has been applied).†

---

* Remember this next time you hear a creationist talking about radiocarbon dating dinosaur bones.

† There is also the added complication that atmospheric carbon-14 has not been constant throughout all of prehistory but has waxed and waned due to various environmental factors. To convert from radiocarbon years to calendar years it is also necessary to calibrate the amount of atmospheric radiocarbon. This can be done accurately using comparison to carbon fixed in tree rings of known date.

One other quirk peculiar to reporting radiocarbon dates is that they are given as years 'before present' or BP. In this case the present does not mean today but AD 1950. There are a couple of reasons for this. Firstly, the initial radiocarbon dates used a bulk standard made in that year, reflecting the natural atmospheric carbon-14 ratio of the time. Secondly, and more importantly, there is now no natural atmospheric radiocarbon level to measure – the legacy of Cold War testing of nuclear fusion bombs on Bikini Atoll and other Pacific islands.* So many tonnes of new radiocarbon and other unstable isotopes were pumped into the atmosphere that the natural cycle has been permanently broken; our planetary legacy writ large in radioactive skies.

In the old days you would measure radiocarbon in your organic samples by turning them into charcoal and sitting beside them with a stopwatch and a Geiger counter to work out the radioactivity levels. These days you can use incredible machines known as accelerated mass spectrometers that are sensitive enough to measure individual atoms in a sample, to work out *the actual number of carbon-14 atoms you have.*

Using radiocarbon dating, British scientists comprehensively sampled more than a hundred cave hyaenas from Eurasia. They struggled to explain the extinction of the species at 31,000 (calendar) years ago. There is nothing in the climate record that could be a cause. Cave hyaenas had survived previous instances of climate change similar to the lead-up to the Late Glacial

---

* In the 43 years between 1945 and 1988, an atomic bomb (either fission or fusion) was detonated on average once every 9.6 days.

Maximum. They had coexisted with Neanderthals (*Homo neanderthalensis*) in Eurasia for hundreds of thousands of years. There is no evidence that any of the hyaena's prey species were under any stress either. Hyaenas still had plenty to eat. It seems like a mystery ... except for one thing: while cave hyaenas had shared Europe with the Neanderthals for scores of millennia, they had never met modern humans (*Homo sapiens*).

Our own species first ventured into Europe about 40,000 years ago and proceeded to oust the Neanderthals from their homeland. To my mind it seems extremely probable that as modern humans took over Eurasia and expanded in range and number, they would have been very active in removing hyaenas from their caves, which were the prime real estate of the time. Our species, with its advanced hunting technologies and differently wired brains, would not have tolerated cave hyaenas squatting in their prospective homes. I can imagine a systematic extermination programme, perhaps as much a part of Palaeolithic culture as fox hunting and tiger shikar were in later times.

It is certainly curious that cave hyaenas are extremely rare in cave art. Maybe this is simply due to their extinction before the full flowering of human expression on the walls of their cave homes. Or it could be that they were deemed vermin not worthy of representation. It's true that the handful of images we do have are not obviously hyaena-like, in stark contrast to the way that the lions, horses, mammoths and other fauna are represented. There is one image in Lascaux Cave in France that could be a hyaena. Or it could be a horse. It looks very giraffe-like, with an elongated neck and sloping back but with the mane, spots and square muzzle of a hyaena. There is another possible example in Chauvet Cave in France, although this one looks more like a cave bear than a hyaena. Its association with *Crocuta crocuta spelaea* is mostly on the basis of spots covering the front of the image. Maybe the artist started painting a bear and playfully added in the spots later to mix things up.

If cave hyaenas did go extinct 31,000 years ago, then it would be hard to explain the images in Chauvet and Lascaux, which are dated to later than this. However, an important lesson we have learned from radiocarbon dating is that the latest bone is not the same as the latest animal. Time and again, the animals of the Pleistocene have surprised scientists by turning up in time periods later than originally thought possible. For cave hyaenas, I wouldn't be at all surprised if they survived later than currently accepted.

Modern spotted hyaenas are hugely complex, unpredictable and intelligent creatures, full of surprises. If we didn't have them as a modern counterpart we would have no way of contextualising the living *Crocuta crocuta spelaea* in all its glory. Using them as an analogue, we can get an insight into such evanescent properties as social structure, dietary choices, life appearance and latrine behaviour. Not every extinct species has such counterparts. In fact, some of the most interesting extinct species get a lot of their allure from the fact that they are unlike anything alive today. When that is the case, scientists have to work a lot harder to tease out the details of their lifestyle. In the next chapter, we meet a predator with no living parallel.

CHAPTER THREE

Sabretooth Cat

Extinct in Britain after
c. 30,000 BC; extinct
globally *c.* 12,000 BC

> The seven teeth which afford the proof of the ancient sojourn of the *Machaerodus* [*Homotherium*] in Great Britain were discovered so long ago as 1826, and their history has been very remarkable [...] There is not the slightest evidence that the *Machaerodus* [*Homotherium*] was more closely allied to the tiger than to any of the other larger Felines, and therefore the very tempting name of 'Sabre-toothed Tiger' must be given up as implying a relationship which does not exist.
>
> William Boyd Dawkins, *The Pleistocene Felidae* (1867)

I'm excited to be writing this chapter because it's about my favourite extinct species of all time. What strikes me as pretty unfair is that I didn't even learn about this genus until I was in my twenties. That's two decades wasted without knowing about the sabretooth cat *Homotherium*. If I had my way, every school would have lessons on biodiversity and extinction. Because how can you expect people to care about the horrific rate at which we are losing species if they don't know the stories of those we have already lost? It's not just dodos and solitaires, moa and mammoths. It's so much more than that. It's complete ways of being that have disappeared forever.

Imagine if time travellers popped in on the history of earth every 10 million or so years after the extinction of the dinosaurs. One or two things would become almost cosily familiar to them on their visits. The diversity of mammals would change between trips, but one particular design would stay in fashion for aeon upon aeon. Starting with the creodonta,[*] but also found in the marsupials,[†] nimravids and barbourofelids,[‡] the sabretooth is a classic evolutionary solution that has popped up again and again, most recently in the Felidae: the cat family. The last felid sabretooths belonged to two species, *Homotherium latidens* (the scimitar-toothed cat) and *Smilodon populator*[§] (the dirk-toothed cat). The Holocene, the geological period we are in now (unless you count the Anthropocene), is actually the first time in many dozens of millions of years that the planet is without a large sabretooth predator of any kind. *Homotherium* and *Smilodon* disappeared with all the other megafauna at the end of the Pleistocene. Geologically speaking, we've just missed them.

Thanks to pop culture (*Ice Age*, *Sinbad and the Eye of the Tiger*, *10,000 BC*, *Walking with Beasts*) many people are

---

[*] Weird carnivorous mammals that were around in the Palaeocene (around 60 million years ago) and not closely related to today's carnivorans but are perhaps relatives of the endangered pangolins (order: Pholidota).

[†] The awesome *Thylacosmilus atrox*, an extinct South American marsupial carnivore with foot-long sabres and a lower jaw that had evolved into an L shape to accommodate bony scabbards for their lethal canines.

[‡] In the order Carnivora but not close relatives of any modern family.

[§] Possibly the most 'metal' binomial ever. Translates as 'knife tooth that brings devastation'.

familiar with the basic sabretooth outline. Enormous canine teeth, hugely muscular body, short tail. It's a pretty good approximation. However, the approximation is a better fit to *Smilodon*. This is the archetypal sabretooth. It is known from nearly a million bones of all age classes – from newborn kitten to toothless senile – from the unique site of Rancho La Brea.

Rancho La Brea is a tar pit found in the middle of downtown Los Angeles in California. Tar seeps up from underground, mixing with mud and hiding underneath small ponds. The tar has been trapping megafauna for over 50,000 years (animals still get caught today), coating them in bitumen so that they might survive as fossils. Rancho La Brea is unusual in that it has very few herbivore fossils and many carnivore fossils. The reason for this is that each unwary vegetarian that got mired attracted a few hunters. Who then also got stuck. And attracted more. And so on and so on until everyone was dead.

Smilodon is a purely American animal, found from California in the north to Tierra del Fuego at the southern tip of South America. But I'm not talking about *Smilodon* here. It has had more than enough people crowing about every aspect of its biology over the last century. *Homotherium latidens* is the deadly apple of my eye. For a start, it was the very first sabretooth cat ever described. The story starts with some very weird serrated canine teeth that were found in a fossil site in Italy (Val d'Arno). Because this site is rich in bear fossils of various kinds, it was initially assumed that the flat, serrated teeth belonged to some kind of bear. So the first researchers, including the famous French naturalist Georges Cuvier,[*] named this supposed beast *Ursus cultridens* and left it at that. Later on, some better material was discovered that included fragments of skull with the weird teeth. These pieces of skull were recognisably cat-like, which posed a problem. Luckily, taxonomists got around this by just erecting a new genus, *Machairodus* ('sword-teeth'), which sidestepped any link to the bears.

At the same time, other fossils of jagged teeth and bits and pieces were popping up in Britain and France. Absolutely central to the history of *Homotherium* is the site of Kents Cavern in Devon. The oldest radiocarbon-dated modern human in Britain was found here. It is a cave rich with bones from Ice Age species. We owe most of our knowledge of Kents Cavern to that curious mixing of theology and archaeology that was so prevalent in the early nineteenth century. Kents Cavern is on land traditionally owned by the Cary family of Torre Abbey. In Georgian Britain, this family, like many others of their class, had a private chaplain, the Reverend John MacEnery, who administered to their spiritual needs and had plenty of spare time for other interests. MacEnery caught a passion for cave hunting. Inspired by the likes of William Buckland, he sought out sites likely to have

---

[*] Cuvier basically invented comparative anatomy: the idea that by looking at any particular bone in a skeleton you could tell a lot about the lifestyle of the living animal. One legend states that he was pranked by a disgruntled student. The student dressed as a devil with horns and cloven feet and burst in upon Cuvier in his bedchamber and declared that he would be devoured. Cuvier gave him a glance and retorted, 'Hmmm, horns, hooves? You're herbivorous! You can't do it!'

'antediluvian' remains. Kents Cavern seemed a likely spot, and one day in 1825 he started digging. It was not long before he was rewarded with fossils. He wrote in his 1859 *Cavern Researches*:

> They were the first fossil teeth I had ever seen, and as I laid my hand on them, relics of extinct (animal) races and witnesses of an order of things which passed away with them, I shrank back involuntarily. Though not insensible to the excitement attending new discoveries, I am not ashamed to own that in the presence of these remains I felt more of awe than joy.

That feeling is very familiar to me, and I hope to you too if you've ever picked up a fossil of your own. Working in the cave in his spare time, MacEnery unearthed cave hyaena, woolly rhino, mammoth and worked flint. His most amazing finds were a cluster of five canines and one incisor of *Homotherium latidens*, the first of this cat in the British fossil record. They caused something of a sensation and MacEnery sent casts of them to the leading fossilists of the time, including William Buckland, Georges Cuvier and Richard Owen, to try and make sense of them.

Richard Owen was perhaps the greatest English anatomist and palaeontologist of the time – founder of the British Museum in South Kensington, coiner of the name 'dinosaur', first to recognise fossil moa bones from New Zealand, polymath and massive shit. Very, very few of his contemporaries had anything good to say about Owen, who seems to have been a ruthless social climber. Charles Darwin, a man so meek that he became physically sick under any emotional stress, had several run-ins with Owen and wrote privately that 'his power of hatred was certainly unsurpassed'. His other enemies are a who's who of Victorian science: Thomas Huxley, Hugh Falconer, Gideon Mantell and untold others. Owen had a particular 'hate-on' for Gideon Mantell, discoverer of the iguanodon, and their lifelong enmity is marvellously described in Deborah Cadbury's *The Dinosaur Hunters*.

Nonetheless, Owen was a superb comparative anatomist, and his 1846 book *A History of British Fossil Mammals, and Birds* is a stone-cold classic with descriptions of many extinct species that stand to this day. Using MacEnery's teeth from Kents Cavern, Owen compared them to the similar ones from Val d'Arno and showed that they were subtly different in size and shape. He put them in a new species, *Machairodus latidens* – the first properly scientifically described sabretooth cat. A British sabretooth. Most importantly, Owen gave an illustration of the canine and incisor he was studying in his work, what we now would call the 'holotype' – the physical ambassador for the species described, allowing all future finds to be compared and contrasted with it. The curved canine ended up in the Royal College of Surgeons in London, where unfortunately it was blown to pieces by the Luftwaffe in the Second World War.

Why was the canine in London? Sadly, Reverend MacEnery died young, at just 43, while he was still writing up his own account of the digging he had done in Kents. As was the way back then, all the fossils MacEnery had collected were sold by auction, sometimes in bulk lots. The unique *Homotherium* teeth were bought by various institutions and ended up in Oxford, London and Exeter. I spent some of my early thirties tracking all these teeth down, and they really are sublime natural wonders. The canines look like something more likely to be found in a shark or a dinosaur.[*]

Because of MacEnery's early death and the haphazard way that he recorded his digging in Kents, many subsequent researchers have been confused about what the hell *Homotherium* teeth were doing in such a late context and how many teeth MacEnery actually found. What makes the teeth from Kents Cavern so out of place is that most of the bones from the site date to within the last 50,000 years. Nearly every other Eurasian fossil of *Homotherium* is 300,000 years old or older. They are teeth out of time. Since the teeth got sold at auction without the associated site details, it has been very hard to account for them all, nearly two centuries after they were discovered. In fact, because Kents Cavern was unique in having such late sabretooth teeth, other Victorian scientists began to privately doubt that the late MacEnery had even found them where he said he had.

To sort out the mess, another cave hunter stepped into the fray. Starting in the 1860s, William Pengelly returned to Kents Cavern with the authority of the British

---

[*] This mistake has actually happened. In East Anglia, where there are Mesozoic strata as well as Pleistocene strata, *Homotherium* teeth and bones have been found, and it takes a bit of technical knowledge to separate the flat, curved, serrated cat teeth from the flat, curved, serrated dinosaur teeth. Basically, mammal teeth have roots and internal dental pulp cavities that are different.

Map 2: UK sites where Homotherium bones have been found.

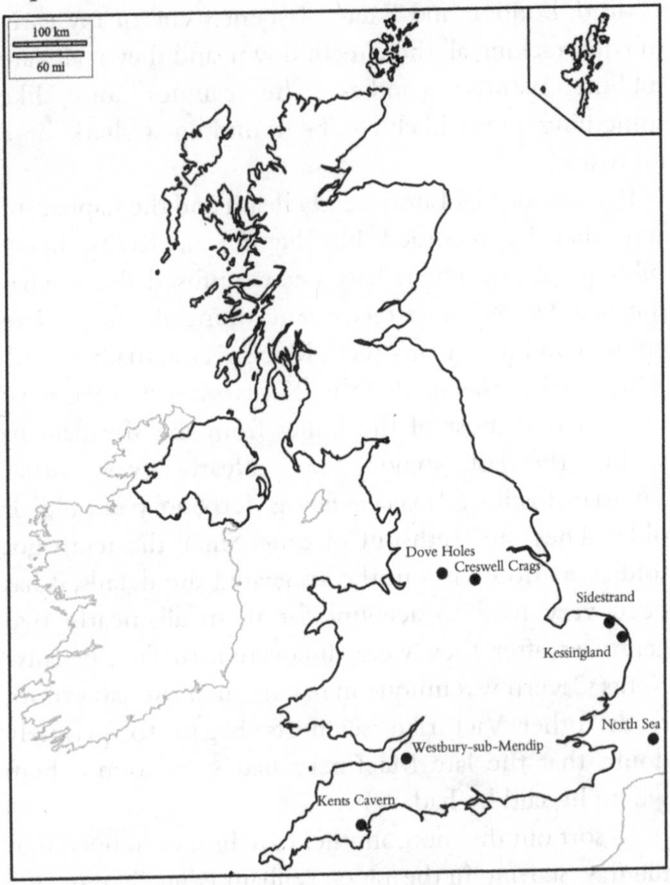

Geological Society, and better excavation techniques, to try and find evidence that MacEnery had been correct. He dug at Kents for eight field seasons between 1864 and 1872 without so much as a sliver of anything *Homotherium*. On his very last day of digging in 1872, he finally struck fossil gold: 'a well-marked incisor of

Machairodus latidens[*] was found on 29 July 1872. The incisor had serrations, perfectly matching the one discovered years earlier by MacEnery. There could now be no doubt that *Homotherium latidens* belonged in the cold earth of Kents Cavern.

Just four years later, *Homotherium* hit the Victorian newsstands again thanks to excavations at the site of Robin Hood Cave in Creswell Crags near Worksop. This complex of caves is pretty famous today since archaeologists recently found Pleistocene engravings of animals on the cave walls – the most northern cave art in Western Europe. Robin Hood Cave was excavated under the auspices of William Boyd Dawkins. A polarising figure, Dawkins was an odd mix of Victorian attitudes. Passionate about education of the working classes, particularly miners, he also seems to have been something of a pompous snob. Certainly, the many labourers under his employ digging in Creswell had little love for him. Despite leaving a legacy of archaeological papers and books that are still consulted today, Dawkins's entire reputation is mixed up with one extremely controversial fossil: a single *Homotherium* canine found in Robin Hood Cave. Whether he was a lucky and fastidious excavator or a fame-hungry hoaxer on a par with the Piltdown jokers depends on one spectacular tooth. It would be hard to find a more literal case of skulduggery.

The canine was unearthed by a workman, in front of Dawkins and several other supervisors, in Robin Hood

[*] Taxonomy is always in flux. Fossil species change their scientific names as more evidence comes to light. What we now know as the bland sounding *Homotherium* (meaning 'same beast' in Greek) is the correct genus for a taxon that has been variously known as *Machaerodus, Machairodus, Epimachairodus, Dinobastis* and more.

Cave on 3 July 1876 at about 2.10 in the afternoon. Robin Hood is a mainly Late Pleistocene site, so it looked like this was the confirmation many had been looking for that scimitar cats had survived into the Late Pleistocene in Britain. Dawkins was unusually jubilant, shouting: 'Hurrah! The *Machairodus*. Oh my! Pengelly will go wild when he hears of this! It will spread like wild-fire over Europe.' For a guy not given to effusiveness or outbursts of passion (especially in front of the lower classes), it was odd. The site supervisor, a Mr Thomas Heath, was certainly suspicious. While Dawkins was out of the cave examining the canine in the sunshine, he looked over the find site. To Heath, it stunk. It looked like somebody had used a crowbar to make a deep hole in the ground and then dropped the tooth down the shaft. Not being of the same social class as Professor Dawkins, Heath kept quiet. The canine was one of the jewels of the excavation and Dawkins wrote up the findings to great acclaim. It was only after these publications came out that the coprolite hit the fan. Heath self-published his own pamphlet suggesting that some of the details in Dawkins's work were wrong and that in his opinion, 'somebody' had planted the tooth in the cave. Things kicked off – big time – and Dawkins and Heath spent the next decade exchanging increasingly angry letters in the national newspapers and scientific journals. Heath stuck to his story that he was certain Dawkins had made many mistakes in his reporting and that the *Homotherium* tooth had been planted. The scientific establishment closed ranks with all the classism that the Victorian period abounded in and Heath was ignored. He conveniently died, aged just 38, and the matter was closed.

So, what actually happened? Did Dawkins plant a scimitar tooth for fortune and glory? Was Heath a troublemaker, resentful of his standing? There are problems with both scenarios. First of all, *Homotherium* canines are not the sort of thing that Dawkins could have picked up at Fortnum & Mason. They are extremely rare; fewer than a dozen are known from the entire British fossil record. Dawkins had contacts in mainland Europe but they are similarly uncommon there. Secondly, Heath's recollections are independently backed up by other excavators who were on site at the time. His written records are clear, and contradict Dawkins in some key details – certainly enough to make us suspicious of Dawkins's memory and motive. There is also a third possibility:[*] taking into account the evidence from both Dawkins and Heath, perhaps the tooth really was found in Creswell but in the spoil heap. Out of context and with no associated stratigraphic information, perhaps Dawkins found it and, to improve its scientific worth, made a wee hole for it and popped it in, making sure that he would be there when it was dug up. This makes sense to me. Dawkins strikes me as someone too full of himself to go to such extreme fakery as getting a rare tooth from another site and planting it. However, he also strikes me as just insecure enough to want to maximise the fame and import of the site he was working on. In this scenario both Dawkins and Heath are sort of right. The tooth does come from Creswell (as Dawkins insisted) but there had been some jiggery-pokery (as Heath was sure). Really, there was no way to resolve this bit of chicanery until relatively recently.

[*] Put forth in Mark White's definitive book on William Boyd Dawkins.

In the archaeological and palaeontological record
one of the major assumptions is that where you find
something is where it comes from. When you dig up
that awesome cave hyaena bone, you assume that it
lived and died where it was found. Things are more
complicated with humans who have the annoying
tendency to be born in one place and die in another. Or
to trade items. This can be really important when you
want to study things like cultural exchange, population
migration or trade networks. Luckily, just like with
radiocarbon dating, science has given us a hack to work
with. When something is buried in the ground it does
not sit there as a static entity. It takes part in a set of
complex interactions with the environment. Over time,
certain elements leach into bones and teeth, while other
elements leach out. Theoretically at least, fossils found
close together within the same site should take up things
like fluorine and uranium from the soil at the same rate.
Therefore it should be straightforward enough to
compare the amount of fluorine and uranium in, for
example, the Creswell *Homotherium* canine and any of
the other bones and teeth from the site to see if they
match. If they don't, then that's some pretty strong
evidence that the canine came from somewhere else.

Luckily, this work has already been done. Back in 1980
samples were taken from the Creswell canine and
Homotherium material from Val d'Arno, France and Kents
Cavern, and were analysed. What the results showed is that
there was no evidence that the Creswell canine had come
from anywhere but Creswell. What's more, the values from
both the Creswell and Kents *Homotherium* suggested that
these were geologically young specimens. They were

probably Late Pleistocene and not Middle Pleistocene fossils, making them the last known representatives of the species in Europe. Still, two controversial sites with no radiocarbon dates[*] and a massive gap between the Middle Pleistocene and Late Pleistocene fossils was not much to go on. And that's where the story went cold for over a century. There were no new finds of Late Pleistocene European scimitar cats since Robin Hood in 1876. Then in 2000, finally things were shaken up a bit …

Fishing the North Sea is a thankless task. Freezing cold year-round, it can kill anyone unlucky enough to fall in within 10 minutes. The fleets working out of ports like Rotterdam trawl the waters for valuable flatfish. Using a technique known as beam trawling, the boats drop a net welded to an enormous metal square. This sinks to the bottom and is dragged behind the ship, with the metal disturbing the fish and causing them to attempt to swim away, only to get captured by the net. As the beam rakes across the sea floor, fish are not the only thing it disturbs. The North Sea is a very recent feature. Until about 8,000 years ago it was open grassland, exposed by lowered sea levels. Mammoths, woolly rhinos, cave lions, bison, Neanderthals,[†] modern humans and other megafauna all called it home. We know this because their fossils are still there beside the buried remains of

---

[*] People have tried to radiocarbon date the *Homotherium* material from Kents Cavern and Creswell but there is very little organic carbon left in the specimens. Not enough to get a good radiocarbon signal. Oh well …

[†] One of the most amazing fossils trawled from the bottom of the North Sea is a chunk of Neanderthal skull complete with diagnostic brow ridges.

river systems, hillscapes and forests. Occasionally the
trawlers will disturb fossils as they scrape across the sea
floor and bring them up to the surface with the flounders
and other flatfish. This can be quite a lucrative by-catch
for the fisherfolk. Many of the fossils make their way
onto eBay or other platforms and can be bought for a
price. Thankfully, Dutch palaeontologists are well aware
of this unusual fishing catch and frequently liaise with
the fisherfolk to examine what they bring ashore.

Back in 2000, a boat brought back a jawbone that
stood out from the fragments of tusks and antlers that
they usually catch. This jawbone could only have
belonged to *Homotherium*. The teeth with their serrations,
the outline of the mandible (jawbone) and its diagnostic
features look like nothing else. It had been dredged from
a region known for Late Pleistocene fossils. Preservation
was good enough that a tooth was sampled for
radiocarbon dating. It came back as just 32,000 years
old – final confirmation that *Homotherium latidens*, the
scimitar-toothed cat, had lived into the Late Pleistocene.
It must have lived at low density between 300,000 and
30,000 years ago to explain the strange lacuna in the
fossil record. Or maybe it had gone regionally extinct
and then immigrated from a different population.
Perhaps future excavators will find more bones to fill in
the gap in the story. Anyway, we can be reasonably
certain that *Homotherium* shared Europe with the people
who painted caves in France and Germany.

In fact, it's quite strange that if scimitar cats were part
of the landscape of the Late Pleistocene, we don't have
any depictions of them in art from that period. Except,
perhaps we do. A small statuette is known from the

Isturitz Cave in the Pyrenees. It is typically feline but with some interesting features. It has a short tail, a massively deep jaw and spots pecked into its side. The legs are long and the body is short. It's not a cave lion (they had long tails). It could conceivably be a lynx (the only other cat with a short tail), but the face looks wrong for a lynx. I guess we will never know for sure what this thing was supposed to be. Especially since the original statuette has been lost.

The best we can do now is reconstruct *Homotherium* from its skeleton. And what a creature it is! Longer-legged and more graceful than *Smilodon*, every proportion of this cat speaks of lithe power. Every tooth in its mouth has serrations (only the posterior edge of the canines in *Smilodon* have such crenulations); its skull is massive and powerfully built. The preserved dimensions of its braincase show it to have a well-developed visual cortex, useful for a hunter active during the day. The neck is immensely powerful, with the bones showing ragged scars from the attachment of colossal muscles.

This ties in to the way it hunted. If one looks at all the works discussing the life of the past, I don't think more ink has been wasted on a single topic than that of how the sabretooth cats used their enormous teeth. Our best guess at the moment is that their bite was something unique – a lost method of hunting that has been called 'the canine shear-bite'. Sabretooths didn't use their teeth to stab, slash or tear (as some have suggested) but evolved a super-precise, hypercontrolled killing bite that probably scythed through blood vessels in the neck, leading to almost instantaneous shock due to loss of blood pressure and near-instant death. When it worked. When it didn't, we see the evidence of broken canine teeth and other serious damage. The canine shear-bite involves the prey being incapacitated first, the predator probably using its muscular front legs to hold it still. Only then does the dread mouth open. Weirdly, the first part to press against the prey is the lower jaw. The comparatively weakly muscled mandible acts as a buttress and pivot point as it nestles against the prey's pulsating neck. Then, *Homotherium*'s monstrous neck muscles come into play and force the upper skull and canines to sweep through the neck like a guillotine. The canines have a unique and very 'gummy' appearance – the sensitive gingiva giving feedback to the brain over how deeply into the flesh the teeth have penetrated, to help avoid breakage. The powerful *Homotherium* neck can then hoist the head back, leaving a massive gaping hole in the prey, showering the cat in arterial blood and giving the hapless meal an impromptu tracheotomy to boot. It must have been a sight to behold.

The canine shear-bite is not just idle speculation. Apart from detailed analyses of the fossil skull, Buckland's modern heirs have indulged in some lovely experimental palaeontology to test it. A few years back some of my colleagues modified a small hydraulic digger to make a facsimile of a sabretooth. The preposterously named 'Robocat' had a resin and metal skull that was an exact cast of a *Smilodon* from Rancho La Brea. Attached to the end of the jib, its articulations matched the living animal and its bite power was supplied by hydraulics. Using Robocat, experiments were performed on bison and deer carcasses to see how much damage the canine shear-bite could do to real flesh. The answer? A lot. The bits of leftover deer and bison were properly mangled.

Such specialised adaptations for hunting imply a specialised diet. We have some clues as to what was on the menu for *Homotherium*: baby mammoth. Our evidence for this comes from another special site. In the middle of Texas, quite near to San Antonio, is Friesenhahn Cave. Inside, excavators found a snapshot of Late Pleistocene Texan life. While many species were present, just two completely dominated the assemblage. There were at least 33 *Homotherium* individuals and at least 34 mammoths based on the teeth and bones found. Included in that sample were the complete and articulated remains of one adult sabretooth and two small kittens. Friesenhahn Cave was probably their den. Close inspection of the mammoth teeth and bones in the cave show another interesting pattern: they are all from juveniles. More specifically, if mammoths aged in the same way as elephants, then they are from young calves between two and four years old. Sounds cute, but even at this young age these mammoths would have been half a tonne to a tonne in weight. That's a lot of meat. Based on this site, it seems that *Homotherium* could have specialised in hunting the young of the mammoths, mastodons, gomphotheres, stegodonts and other proboscideans it shared its global range with.

There is one particular aspect about the lives of young mammoths that may explain their susceptibility to this sabretooth. In elephants today, when they are around two to four years old, they first start to gain their independence from their mothers. These giant toddlers begin to forage away from the eyeline of the other herd members. It is a dangerous time. The unique lion population of Chobe National Park in Botswana occasionally tackle elephants. When they do, it is always young animals under the age of 11. It takes only a little imagination to picture one or a few *Homotherium* slinking around a young mammoth that has wandered off to explore some new smells. Almost without realising what is happening, the calf is knocked over by a sudden charge from a sabrecat and pinned down by those bear-like legs. Quickly, the teeth are flashed and the confusion ends with a bonanza of meat for the cat and for the kittens waiting patiently in the den. Any human must have been inspired and/or terrified by seeing *Homotherium latidens* take down its prey on the plains of Europe and North America.

But let's circle back to those canine teeth from Kents, because they have one more secret to reveal. While the analyses of fluorine and uranium in the Kents canines I mentioned earlier do show that the teeth were from the Late Pleistocene, and had been in Kents for a long time, that's not the whole story. Recent work from 2013 used even finer tools to probe the enigma of these teeth. Teeth contain the whole history of their owner wrapped up in enamel and dentine. They can also show their life after death. Sectioning teeth from Kents Cavern, splitting them right across, scientists used lasers to zap miniscule amounts of tooth for study. They sampled at intervals from the outside of the tooth right to the inside of the root. And this is where the surprise comes in. While the relative amounts of fluorine and uranium on the surface of the *Homotherium* tooth matched the local environment of Kents Cavern, showing it had not been recently planted, the inside of the tooth said something different. The internal bits of the Kents *Homotherium* canine had ten to a hundred times the amount of uranium that was found on the surface of the tooth. What that means is that the Kents *Homotherium* teeth must have originally been somewhere else before making their way to Devon. But that movement *must have happened in the Late Pleistocene*. What this research shows is that Ice Age people were moving *Homotherium* teeth about. They were trading them! Or at the least attaching some kind of cultural value to them and carrying them from place to place. This makes more sense when you know that a few people have pointed to the Creswell canine showing signs of being turned into a pendant.[*] There is even a supposed example of a *Homotherium* canine fossil found in an Etruscan tomb and kept at a museum in Florence. The teeth are so impressive that it's not surprising they have appealed to people through the ages.

And it perhaps goes back even further than we might think. The Middle Pleistocene site of Schöningen in Germany is famous for the discovery of eight wooden spears. About 2 metres (6.5 feet) long and made of supple spruce and pine, these were wielded by *Homo heidelbergensis* nearly half a million years ago. They've been preserved by anaerobic conditions at the site. The humans who lived here 300,000 years ago hunted extinct Mosbach horses and other mammals, and the site is full of their bones along with the spears and flints used to kill

---

[*] No one has looked at this fascinating idea in any depth. It was first told to me by Andy Currant at the National History Museum in London, who ascribed it to Roger Jacobi, although William Boyd Dawkins mentions something similar in one of his papers. The Creswell tooth does look a bit like it has been modified at the root by some flint tools. Looking at it under an electron microscope could maybe resolve this.

and process them. In amongst the herbivore remains are also a handful of glossy black* *Homotherium* teeth. Schöningen shows that our early cousins were able to plan complex hunts of wild animals in difficult landscapes using tools of wood and stone. It used to be thought that sabretooths and hominids shared a 'special relationship' whereby the leftover bones of a sabrecat hunt provided access to marrow and carrion. At Schöningen it seems this is not the case. There are actual horse bones from here that have both cut marks (from flint tools) and tooth marks from carnivores (like *Homotherium*). In all cases, the bites have occurred over the top of the cut marks, showing that people had first access to the bones. *Homo heidelbergensis*, not *Homotherium latidens*, was top cat here. As if to emphasise this point even further, there is one bone from Schöningen that I think summarises the relationship between early humans and sabretooth cats better than any other. It is a humerus (arm) bone from *Homotherium*, and it has been used as a tool to knap flints. If Middle Pleistocene material culture was anything like today's, then this could have been the Rolex of knapping tools. Nothing shows off your status as top predator better than using the bones of the previous top predator to make your tools. You could say that our relationship with the magnificent beast was probably antagonistic, at best.

So, where do we stand in our understanding of *Homotherium* now? The one key question that I've done my utmost to contribute to is that of the phylogeny of the sabretooth cats. Where do these species sit within the

---

* Stained by their surroundings. Schöningen is a lignite mine – excavated for the coal-like mineral.

myriad grandeur of the carnivores? We can see from the bones that *Smilodon* and *Homotherium* are allied to the cats, but how closely? Are they actually cats? Or are they something cat-like? You can think of it like the giant panda: is it part of the bear family or just outside, looking in? My very first piece of published research looked at this question in *Smilodon*.

My supervisor had heard about bones from a cold cave site[*] at the southern tip of Patagonia, which were stored in a collection in Amsterdam. The cold, dry conditions of the cave led to excellent preservation, and the bones of *Smilodon populator* were in amazing shape. He sampled small chunks of the bone, and one of my first tasks as a scientist was to process the tiny pieces and extract their DNA. We managed to get about 1,000 base pairs (the chemical letters in the alphabet of DNA) from them. In comparison, the human genome project sequenced nearly all of the 3.6 billion base pairs that make a human genome. Still, I felt pretty proud of myself as our sabretooth bones were over 11,000 years old.

Using this DNA sequence and comparing it to equivalent sequences from lions, tigers, domestic cats, hyaenas and mongooses allowed us to create a family tree for *Smilodon*. It's exactly the same kind of analysis that gets done for DNA fingerprinting. We found that *Smilodon* (and by implication the rest of the sabretooth

---

[*] This was Cueva del Milodón, or Mylodon Cave, semi-famous for the discovery of the preserved skin of the giant sloth, *Mylodon darwinii*. The beautiful preservation of the skin (with copious red hair and even small bones embedded within like chain mail) was a nineteenth-century sensation. Its apparent freshness encouraged the *Daily Express* to send a reporter to look for living ground sloths. He didn't find any.

subfamily) are about equally distant from all modern cats on a branch to themselves somewhere between the cats and the hyaenas. This was a great answer, giving me my first publication and getting a reasonable amount of press in the national and international media – very gratifying for the culmination of three years' work.

The one thing that I regretted at the time was that we couldn't get any good DNA from *Homotherium* to include in the analyses. We had tried pretty hard, though. I extracted DNA from many samples of *Homotherium* from all over North America but none of them had enough usable sequence for comparison. My love of *Homotherium* was unrequited. Nonetheless, I put that to the back of my mind and concentrated on other things for the next few years. The issue of *Homotherium*'s phylogeny had to wait.

A decade later I was working on the genetics of lions in Copenhagen. A colleague had sent me some bones from Beringian cave lions (*Panthera spelaea*) and I was working hard on using new techniques and cutting-edge facilities to get the DNA out. The bones had come fresh from the permafrost and had been on dry ice on their trip over from Yukon to Denmark. Now, instead of trying to get small, precise fragments of DNA and stitching them together, we had the technology to amplify billions of strands of DNA and then use massive computing power to fish out those pieces belonging to the bone. This 'next-generation' technology is fantastic in terms of producing DNA data from old bones, but it now requires that biologists have as much familiarity with programming languages and user-unfriendly open-source specialist software as they should with the natural world. Personally, I much prefer spending time in the lab.

One of the lion samples I had been sent from Canada proved a bit unusual when it came to analysing the data. I had blagged some basic skills in manipulating the next-generation DNA data, and when I was looking at the genome of what should have been a cave lion, something was weird. I focused on a small section of DNA that I was very familiar with (from spending hours upon end staring at a computer screen filled with DNA sequences). Something was wrong. There was a gap. In a living lion this would have been a death sentence, as the sequence would produce a mutant protein. I recognised it from somewhere else, though. My colleague and friend Professor Beth Shapiro, who was working at the University of California, had just published a paper that had a small snippet of DNA from a Minnesotan *Homotherium* bone. They matched. This wasn't an ordinary cave lion bone. It was something much, much rarer and much, much better. We had a permafrost *Homotherium* bone that had good DNA inside it! In the decade since I first worked on *Smilodon* DNA, the sequencing technology had improved so much that we now had a real chance of generating a lot of data. And the more data, the more certain we can be of the species tree.

Now seems as good a time as any to go into some detail of how we work with DNA from long-extinct species. Sadly, it's not the case that you can just fling some chemicals at the bone and get the DNA sequence back. When bones have been in the ground for 50,000 years they get chomped up by entropy, by bacteria and fungi. Usually, nothing survives. Occasionally, a few precious picograms of DNA will hang on. We can't do anything with this. The keystone technique of all ancient DNA work, and the only way we have been able to look at the genetics of cave hyaenas, *Smilodon* and *Homotherium*, is something called the polymerase chain reaction, or PCR[*] for short.

PCR works by taking all the ingredients that replicate DNA within a living cell and transplanting them into a test tube. When cells divide, the magic of the double helix allows it to split into two complementary single strands that can then be built back up using raw materials within the cell. Watson and Crick[†] often get the credit for discovering DNA, but actually scientists had identified DNA a century before. Watson and Crick worked out the *structure* of DNA, and it is the structure that caused Crick to joyfully exclaim to a packed Cambridge pub that he had 'found the secret of life'. In DNA, there are only four 'letters' (A, C, G and T) that make up the entire language. As speech, it is not easy on the ears, but as a code it is superb. Within the two strands that make up the double helix there are only two rules: A must bind across the strands with T, and G must bind across the strands with C. This means that either strand has all the information necessary to reconstruct the whole double helix. It's kind of comparable to a palindrome. If I wanted to say to someone the palindrome 'madam, I'm Adam', I could give one person 'madam, I' and the other 'I'm Adam' and both would be able to

---

[*] CPR is when you need to shock your heart into working; PCR is where your heart is shocked that nothing works.
[†] And of course Rosalind Franklin too, who did a lot of the work Watson and Crick used for their paper and then was denied a Nobel Prize by dying young.

reconstruct the whole message.* In a similar way, during cell replication the double helix peels apart, the complementary 'letters' are added to each template strand and voila – two new double helices!

PCR does all this in the test tube. You start with the picograms of template DNA extracted from your bone and add in lots of free As, Cs, Gs and Ts and a few other reagents and then you include the special enzyme that replicates DNA inside cells: DNA polymerase. PCR uses a special polymerase that comes from a particular organism. *Taq* polymerase is extracted from the extremophile *Thermus aquaticus*, a bacterium from the hot springs of Yellowstone National Park. What is unique about *Taq* polymerase is that it works at about 70°C (158°F). Human DNA polymerase works at 37°C (99°F). How PCR works in the test tube is because of this temperature specificity of *Taq*. When *Taq* is mixed into your test tube with your DNA and other reagents, you can heat it up to 95°C (203°F) and all the double helices of DNA will unzip into single strands. Cool it down to 40°C (104°F) and the single strands will try to pair up again. Now heat it up to 72°C (162°F) and *Taq* will happily fill in all the gaps to produce two copies of the double helix, where before there was one. Here is the clever bit.† Each time you cycle through those three temperature steps, the new DNA you produce can act as a template for the next cycle. What this means is that after one cycle you have twice as much DNA as you

---

* If you don't believe me, see if you can work out the whole sentence from this half-palindrome: 'lid off a d …'.
† The inventor of PCR, Kary Mullis, indeed received the Nobel Prize in Chemistry for his invention.

started with. After two cycles you have four times as much. After three cycles you have eight times as much. Do this for 35 cycles and you have more than a billion copies per starting template. A billion here, a billion there, and pretty soon you're talking about real DNA. These are the numbers of molecules we need to have in the lab to do any downstream work with. PCR is crucial to doing this.

When I realised that our cave lion bone was a *Homotherium* bone, I was pretty ecstatic. We used the power of PCR and next-generation sequencing technologies to massively amplify all of the DNA we had extracted from the bone. Then, with some bioinformatics know-how, we stitched together the complete mitochondrial genome of our Yukon *Homotherium*. Because the field of ancient DNA is small, we were able to collaborate with some colleagues in Germany who had access to that famous North Sea *Homotherium*. They used our data as a starting point to fish out the sabretooth DNA from their sample. Together, we were able to publish a definitive family tree for the sabretooth cats. What we saw was that *Smilodon* and *Homotherium* were each other's closest relatives, sitting on a branch of the evolutionary tree just outside the cat family. They had both diverged from the modern cats about 20 million years ago. What was interesting, though, was that *Homotherium* had diverged again from *Smilodon* about 18 million years ago. Despite being lumped together as 'sabretooth cats', the species were very different. One other neat finding was that the Yukon *Homotherium* and North Sea *Homotherium* were actually practically identical. Genetically, they were much closer to each other than, say, a tiger from India and a tiger from Siberia. Previously, European and American *Homotherium* were split into the species *Homotherium latidens* and *Homotherium serum*. The genetic data seems to show that actually they weren't different enough to be distinct species. Since Richard Owen's description of the Kents material is the oldest, all *Homotherium* specimens from England to Canada should probably be known as just *Homotherium latidens*. Ironically enough, our genetic investigation of this incredible sabretooth had, on paper at least, caused the extinction of *Homotherium serum*.

Homotherium has always been an enigma; a mystery that brought out the best and worst in scientists, making and breaking the reputations of those who worked on it. I like to imagine it as master of the Pleistocene steppe, gracefully wandering, shoulders lolling, as it scans the tundra for its next meal. As a hunter it must have been awe-inspiring. But, incredibly, it wasn't the biggest cat around. Sharing its territory was another giant of the Ice Age. A cat that truly was lord of the plains and king of the tundra: the cave lion.

CHAPTER FOUR

Cave Lion

Extinct in Britain after
c. 38,500 BC; extinct
globally *c.* 12,000 BC

> The first evidence of the discovery of *Felis spelaea* is afforded by a figure of an unequal phalange, appended to a paper on the Dragons of the Carpathians, written by Dr John Hain in 1672.
>
> William Boyd Dawkins, *The Pleistocene Felidae* (1867)

Britain has its own native felid. The Highland tiger is a runt of a thing resembling a floofy domestic tabby. However, as many have discovered to their cost, this little ball of venom and spite has claws. The motto of Clan MacPherson, 'Touch not the cat bot a glove', shows that Highlanders are well acquainted with the fury of wildcats. It's in trouble too. While we are happy to throw money at conservation for real tigers in India and Indonesia, our own endemic cat is rapidly disappearing. Sucked down and down into an extinction vortex. Pushed there by gamekeeping persecution, and riding the eddies through hybridisation with feral domestics. The wild glens and straths cannot even support a 5-kilogram (11-pound) wildcat thanks to humans. It will, in all likelihood, soon be gone. But we won't notice for a little while. The abundance of feral cats and hybrids will ensure that the true wildcat will disappear without fanfare, masked by the abundance of a similar form. This will be the first extinction of a felid species in Britain for a millennium, and only the second felid extinction since the end of the Ice Age.

It used to be thought that Britain was home to actual tigers. I'm not talking about the escaped exotics and alien big cats that are reported from Bodmin to Buchan.[*] When Buckland was excavating Kirkdale (see Chapter 2), he reported teeth and bones of tigers mixed in with those of rhinos, mammoths and hippos.[†] Of course, it's extremely difficult to identify large felids with their skins off, and a tiger skeleton is very much like those of other big cats. It's forgivable that in this small detail Buckland was actually wrong. While Britain has been home to cheetahs, European jaguars and leopards, it has never had tigers. What Buckland had found in Kirkdale, and what has been found all over England and Wales (and even beneath the North Sea), was something quite special: the cave lion, *Panthera spelaea*.

I've been obsessed with cave lions for a long time, and they have slowly overtaken a large percentage of my research output. I, and colleagues in the field, have

---

[*] And why are these exotics never reported as tigers but always as 'black panthers'? There is no such cat as a panther – a personal bugbear of mine. 'Panther' can refer to the cougar, which is light-coloured, or a rare black form of the leopard or jaguar that is occasionally found. Tigers are probably the most common big cat in captivity (both legally in zoos and illegally in private ownership), and yet you never hear an eyewitness say they saw a tiger. Perhaps their stripy camouflage is so good that the only people who do see them never return to file their sighting …

[†] The most northerly British record of *Hippopotamus amphibius*. I like to think of them oozing about in the Ouse.

Map 3: UK sites where cave lion bones have been found, along with calibrated radiocarbon dates where they are known.

looked at their genetics in detail, teasing out nuggets of genetic gold from their long-dead bones. Despite going suddenly extinct approximately 14,000 years ago, cave lions were probably *the* most successful land mammal before humans came along. Their range was unbelievable. From southern Spain to Yukon, Canada, by way of

Britain,* Siberia and everywhere in between. There was also a related population of lions, descended from cave lions, in the continental United States and as far south as Mexico. When you factor in the ancestors of our modern lions in Africa, the Middle East and South Asia, you had a species complex that spanned almost the entire land mass available to placental mammals. Not until modern humans entered the Americas, Europe and Australia during the Late Pleistocene did any other species' range surpass that of the lion.

Their discovery happened during the Enlightenment, when curious polymaths put away the idea of dragons in caves and started using reason to work out what the strange bones they kept finding actually were. While Buckland, MacEnery and Pengelly were working in the caves of England, Germany had its own illustrious sons digging away. One of these men was Georg August Goldfuss, a professor of zoology and mineralogy in Bonn. Amongst his many successes, including being the first to scientifically name the koala, he identified some remains from Zoolithen Cave in Bavaria as an extinct type of cat, *Felis spelaea* (literally, 'cat of the cave'), based on detailed analysis of a skull. And it's a pretty incredible skull. If you're used to seeing skulls on display in museums that have been expertly prepared by researchers to, well, bone-white perfection by the use of dermestid beetles or liberal dousing with bleach, then it's difficult to ever appreciate palaeontological

---

* Landseer's bronze lions in Trafalgar Square quietly hide the secret that real cave lion bones were found underneath when the area was first excavated. Like the cave lion, the Barbary lions that acted as models for Landseer's vision are also now extinct.

material in the same way. It's often scrappy, or damaged, or a particularly unpleasant shade of grey or brown thanks to millennia under the ground. It's rare to find anything as beautiful as the rich, chocolatey browns you find in the tar pit bones of Rancho La Brea, or the dark, bark-like patina you see on bones hoisted from the bottom of the North Sea.

The cave lion holotype skull unearthed by Goldfuss is a sickly yellow, and it has seen some serious action. Right in the middle of its sagittal crest (the bony line you can feel running front to back if you rub a cat's head) there is a chunk of bone missing. Something took a bite out of this lion's head – while it was alive. We know this because some of the bone has grown back. Whether this was intraspecific fighting between rival males (as happens in modern lions), or due to an altercation with a cave hyaena, cave bear or some other member of the European Pleistocene, we don't know. But the skull is in good enough condition that Goldfuss was able to compare it to African lions and Indian tigers, and confidently declare that it was something else.

But how different was it? Was it a European relative of the lion or the tiger? Getting an answer to that question would be difficult, and different researchers have come to very different conclusions over the past two centuries. Some have sided with Buckland and called it the cave tiger (*Panthera tigris spelaea*). Some have placed it as a subspecies of modern lion (*Panthera leo spelaea*). But most have converged on the idea that it was lion-like but not exactly a lion. We can think about it using the analogous situation of the beautiful clouded leopards (*Neofelis* spp.*). Until just a few years ago these incomparably beautiful† cats were lumped together as one species. It was only with the advent of molecular biology and comparative DNA analysis that researchers discovered that the animals in mainland South Asia were distinct enough from the animals in the Sunda Islands to warrant separation: *Neofelis nebulosa* was split into *Neofelis nebulosa* and *Neofelis diardi* respectively. Using the DNA evidence as a prompt for further study, differences were then found in pelage, skull shape and dentition. In fact, the two species may have been separated for over a million years, plenty of time to evolve substantial differences in behaviour and ecology. For the cave lions we also have plenty of DNA evidence for their separation (which I spent a couple of years of my PhD painstakingly extracting), and strong skeletal evidence for their alliance with the modern lions. As for their behaviour and ecology, this is difficult to gauge for any extinct species (obviously), but we are incredibly lucky to have some very suggestive lines of evidence to look to.

For instance, the constantly frozen ground of Siberia and Alaska contains many complete mummies of

---

* The taxonomic shorthand 'spp.' refers to multiple species within a genus, whereas 'sp.' means there is only one referred species in the genus.
† Their lovely cloud-shaped markings that so distinctively identify this taxon are described in the species name *nebulosa*. Clouded leopards famously have enormous canine teeth that are proportionately larger than those of any other cat species, bordering on the ratios found in extinct sabretooth cats. Despite this, as far as I know there is only one recorded instance of an attack by a wild clouded leopard on a human and it was non-fatal.

mammoths, rhinos, bison, horses and other extinct creatures. A bison mummy from Alaska nicknamed 'Blue Babe' (more on this in Chapter 7) shows strong evidence of being hunted by cave lions in the claw and teeth marks left in its skin. Russian hunters have also recently discovered two perfectly preserved cave lion cubs eroding out of the banks of the Uyandina River in Siberia.* In time, perhaps they will tell us as much about cave lions as Tutankhamun has told us about ancient Egypt.

There are indirect ways to look at diet in cave lions too. Stable isotopes are a way of assessing dietary webs in fossil communities. Unlike unstable radioisotopes (like carbon-14 that decay over time), stable isotopes are found in fixed percentages in the atmosphere but, importantly, are incorporated differently into living tissue, depending on certain factors. Two stable isotopes commonly used in palaeontology are carbon-13 and nitrogen-15. Plants can vary in their carbon-13 values depending on whether they grow in marine environments, or use certain methods of metabolising carbon dioxide during photosynthesis. Luckily, grasses and most trees use different pathways (called C_4 and C_3 respectively), which allow for the differentiation of strict browsers and grazers according to the amount of carbon-13 in their tissue. Using stable isotopes in the fossil record it can be easy to tell a grazer (e.g. sheep) from a browser (e.g. giraffe) from a mixed feeder (e.g. deer), and this difference in isotopic values is passed on to the carnivores that eat

---

* Two more cave lion cubs have also just been found. One has been named Spartak, the other Boris. You wait 14,000 years for a cave lion and then four turn up at once!

them. Using carbon-13 you can tell whether a carnivore has specialised in grazers, browsers or mixed feeders. Nitrogen-15 works in a similar way, but because it is mostly consumed via protein, it gets enriched as you travel up the food pyramid, with omnivores and carnivores showing higher values than herbivores. Naturally, in fossil food webs you have to test the values of all potential predators and prey from similar time points. This has been done for cave lions with interesting results. In Europe, cave lions from Belgium and Germany show very individual prey preferences. Some seem to have preferred young, juicy cave bear cubs,[*] while others specialised in hunting mature reindeer.[†] This makes sense in light of the hunting strategies seen in modern lions. Some African prides (e.g. those in Chobe, Botswana) can successfully tackle elephants or giraffes, while for the most part they take whatever is going, which is usually mid-sized ungulates like antelopes, wildebeest, warthogs and zebras. Lions are smart enough to know that what makes a good meal is whatever can be caught easily. Their coordinated, social hunts are one of the natural world's most impressive sights.

It's unfortunate that no modern field biologists ever got a chance to study the cave lion; however, we have something almost as good. In the caves and grottoes of the Old World, Pleistocene people expressed themselves

---

[*] Cave bears, being strictly vegetarian (as determined by their stable isotope values), probably tasted pretty good. Their propensity to hibernate in caves would have made them easy pickings.
[†] Reindeer are the same species as caribou. If you want to get technical, reindeer are semi-domesticated caribou. During the Ice Age they were plentiful all over Europe.

in ochre and grease, leaving enigmatic pictures in the hallowed spaces of Lascaux, Chauvet and Les Trois Frères, and even carved in antler and ivory. Anyone who has looked at the monumental panoramas in Chauvet Cave knows that these people were phenomenal observers.* They lived in an intimate embrace with their environment and the species that shared their lives. In essence, the art they left behind acts as a literature of the vibrant life that surrounded them.

One of my favourite examples comes from the Pyrenean site of La Vache. Here, during the Magdalenian period (around 15,000 years ago), an artist of uncommon skill used a flint point to carve a row of three felids onto a piece of rib. The image is damaged and cracked and it's difficult to tell from the heads whether they're supposed to be lynxes, leopards, sabretooths or cave lions. Except for one thing. Delicately traced are the long, flowing tails (thereby excluding the bobtailed lynx and sabretooth), which have a tufted flourish. Only lions have this feature. We may not be able to interpret all the nuances of what they were trying to tell us, but there is information there to be extracted. Cave lions are the most numerous carnivores depicted in Pleistocene art; in Chauvet, La Vache and other places they are often shown in dynamic prides of multiple individuals, sometimes gazing longingly at the bison and horses painted nearby. Waiting 30,000 years for a hunt that will never be completed.

---

* Picasso was supposed to have said 'we have learned nothing in 12,000 years' after a personal visit to Lascaux Cave. This apocryphal story is repeated a lot, despite the fact that Picasso never visited Lascaux.

Already, we learn something about this extinct species: like their modern counterparts, the cave lions lived in prides. It's good to take a minute to recognise how important this information is. Unlike every other one of the 37 species of felid, lions are the only truly social cat. Their nearest relatives, leopards and jaguars, are ferociously solitary. Our prehistoric ancestors have imparted some knowledge down through the millennia that we could never have gleaned from the bones alone.

It goes deeper than this too. The male lion, like many other carnivorans, has a rather, well, *noticeable* external scrotum. One spectacular image from Chauvet has two lions side by side, a conspicuous male and female pair, judging by their outlines. The male, with his obvious scrotum, is engaging in a behavioural routine known from modern lions as 'hunkering down'. Here, the usually much larger male bends his front legs to bring his head in line with the smaller female, as a prelude to courtship. And that is when this one image betrays a second secret. The adult male in this coupling has no mane. In all the dozens of images of cave lions from Ice Age art, there are no manes to be seen at all. At best, there is the faint suggestion of a facial ruff, akin to that seen in tigers' fluffy cheeks. Despite how inextricably we link modern lion 'maleness' to the halo of hair that only they possess, the cave lion went a different way. The best explanation we have for why this is comes from a deeper understanding of the evolutionary history of the lion group.

Lions are the sister species of leopards, with jaguars probably being their next closest relative. It's easy to see this close relationship if you look at the cubs: all the little fuzzballs are pretty identical at first and only later do lions lose the rosettes* that they are born with. Like us, the first lions probably evolved in Africa. The fossil record is patchy, probably due to the habitat the animals used not being conducive to fossilisation. The earliest bones that we can confidently ascribe to lion-like cats are found at the site of Laetoli in Tanzania and dated to the Late Pliocene (three to four million years ago). Laetoli is rightly famous for the impressions left in volcanic ash by the footprints of our hominid ancestors (probably *Australopithecus afarensis*, the species typified by 'Lucy') but also has a comparatively rich fossil record. The next fossil bones we can assign to lions are only 1.4 to 1.3 million years old and also come from another famous Tanzanian hominid site: Olduvai. These fossil lions are big – bigger than any of the lions we have today. This trend to largeness reaches its peak with the species known as the ancestral cave lion, *Panthera leo fossilis*. This giant cat (perhaps the largest ever to evolve) is actually reasonably well known from fossils found in Europe from around 700,000 years ago. Like us, lions seem to have suffered from wanderlust and repeatedly left Africa to see the world. This may not be entirely a coincidence. The ancestral cave lion left Africa around about the same

---

* Not spots. Cheetahs and servals are the only truly spotted cats. Lion cubs have rosettes, which are complicated clusters of dots arranged in a circular pattern. Jaguars and leopards also have rosettes. In fact, a good way to identify leopards and jaguars (aside from geography) is by their rosettes. Jaguar rosettes tend to be larger and have a central spot with other dots arranged radially around this; leopards have no central spot.

time as the founders of the Neanderthal lineage. Changing climate during this time period opened up routes out of Africa and our meat-loving cousins may actually have followed lions by choice, taking advantage of the plentiful marrow and other scraps left behind from their hunts.* This joint diaspora led to the Neanderthals and the cave lion occupying similar territories in Eurasia† – sometimes even occupying the same cave systems (though probably not at the same time).

So, numerous lines of evidence lead us to the idea that fairly early lions split into two groups: one leading to our modern maned lion, and the other to the cave lion (via *fossilis*). In studies of *fossilis* and *spelaea*, there are clear signs of marked sexual size dimorphism, as well as the cave art showing pride structure. This leads us to think that group living evolved very soon after their split from the ancestors of leopards, allowing for more efficient hunting, a greater availability of calories (i.e. larger prey) and a chance to grow much bigger than their forebears. In the stay-at-home lineage of lions that remained in Africa, along with size dimorphism, at some point the

* Some modern human groups still preferentially scavenge lion kills, for example the Ogiek of Kenya. Unlike hyaenas and some canids, felids can only really consume soft tissue due to the shape of their teeth, and thus leave large bones, and their marrowy goodness, completely untouched.
† There is also evidence in both humans and lions for a second shared exodus. In the Late Pleistocene, anatomically modern humans left their ancestral homeland again and spread across the globe, leading to the current worldwide distribution of humanity. In lions, this second exodus produced the populations that until very recently occupied the Middle East and South Asia.

males seem to have picked up the mane trait, and the selective advantage was such that it was passed down to all the modern lion groups that flourished in Africa and Asia.

Manes themselves are a bit of an enigma. Obviously, they look impressive (both to us and to other lions) and give the owner the advantage of looking bigger – important when male–male aggression is a constant threat. They also have more occult benefits, and have been the subject of a lot of discussion. Fieldwork has shown that lionesses find darker manes sexier (and colour is intimately linked to testosterone levels). This was hilariously tested in the field by building fake lions with variously shaded tints of bouffant and leaving them in pride territory, while noting which ones got the most attention from passing females. People as eminent as Darwin have suggested that manes protect from injury while passing through the thorny veld, or protect the vulnerable throat and neck from damage during fighting. These protective roles seem to be only secondary to their obvious sexiness. There is also the fact that manes respond to temperature, growing lusher in cold climates and sparser in warmer ones. The last surviving Asian lions of the Gir Peninsula in India tend to have scraggly manes at best. Similarly, the lions of Tsavo* in Kenya have the barest bumfluff despite doing everything that male lions in other populations do. Thus, despite being nearly

* Home of the 'Tsavo man-eaters' who ate their way through 135 railway workers of the Kenya–Uganda railway in 1898 before Lieutenant Colonel J. H. Patterson shot them both. They are now mounted in the Chicago Field Museum, and the story was the basis for the film *The Ghost and the Darkness*.

ubiquitous in modern lions, the mane itself is not absolutely essential and we can see that cave lions could have easily done without them.

We can now picture the cave lions of Britain and Europe as large, superficially lion-like cats, furry but maneless, living in small prides, making their home in the tundra-steppe, where they occasionally would have overlapped with humans. How did these people interact with them? Disinterested observation wasn't an option. Think about how much the modern lion has shaped culture in the English-speaking world, appearing on flags and coats of arms or in fairy stories, and popping up in many distinct guises in our language. Leonine, lion-hearted, lionised, into the lion's den, the lion's share ... We have some compelling evidence that our naked obsession with the king of beasts* started before we were fully human.

One of the earliest signs of direct contact between people and lions is from northern Spain. Here, at the site of Atapuerca (again, another hominid site, famous for finds of *Homo antecessor* and *Homo heidelbergensis*), we have evidence of two very different kinds of interaction. First, there are the hominid bones from Sima de los Huesos ('Pit of Bones') at Atapuerca, where of 1,600 analysed, many show clear evidence of gnawing by cave lions. The scratches and striations are such that

---

* Not the king of the jungle. Lions tend to shun dense vegetation and no population has made their home in the rainforest. In fact, the biggest split within lineages of modern lions is that caused by the inability of lions to cross the equatorial rainforests in Central Africa.

they exclude hyaenas, bears or other large carnivores and must have been made when fresh by lions. It's possible that the cave lions dragged the bodies into the cave to eat them. Later they were buried deeper by mudflow.

Conversely, at the nearby Gran Dolina site in Atapuerca, the remains of an ancestral cave lion were found with cut marks showing how it was defleshed and gutted before being filleted. Even the bones were broken for their marrow. Both these sites are slightly under a million years old. Hunt or be hunted. Kill or be killed.

Fast-forward to the end of the Ice Age and we see another burst of data. After our Neanderthal cousins had gone extinct, there is evidence that modern humans were using cave lion bodies in interesting ways. From sites in Germany and France there are cave lion incisors and canines (hugely impressive-looking teeth, by anyone's estimation) that have been turned into pendants for necklaces (as has been suggested in Chapter 3 for *Homotherium*). Were these scavenged or hunted? We don't know. There has also been a recent find in northern Spain of a cave lion rug. Yep. The cave site of La Garma preserves the cut-marked remains of many lion phalanges in a position suggesting that they once belonged to a complete pelt. Surrounded by large rocks, it immediately suggests to me a cosy, luxurious bed.

However, the trump card for human infatuation with cave lions was found in south-western Germany, in a cave near Stuttgart. For my money, it is the most beautiful and enchanting object ever dug up by an archaeologist and has the potential for almost endless speculation about its meaning. This is of course the enigmatic *Löwenmensch* ('lion-man') of Hohlenstein-Stadel – a figurine of pure mammoth ivory, whittled by expert hands into a fantastic chimera of man and cave lion. Standing about a foot tall on its human legs, it has the front paws and head of a lion, with rounded ears, a questioning smile and sightless eyes. Thought to be about 40,000 years old, it is, by a long shot, the oldest evidence we have, anywhere, for the infinite capacity for human imagination to simply make things up. Experts painstakingly pieced it back together from 632 fragments unearthed piecemeal over the past century. Can you imagine? A three-dimensional jigsaw of that magnitude, with no box to crib, gradually forming into a fever-dream image. What the hell does it mean? It's very tempting to slap the label of religion onto it and quickly move away from its uncanny weirdness. However, I think that in this case it's the only explanation that makes a jot of sense. Especially when you consider that there are actually multiple *Löwenmensch* floating about. A second, smaller figurine was unearthed in another nearby cave. Whatever the lion-man meant, the meaning wasn't restricted to one small group. This was an idea that travelled. It had cachet. Did people worship cave lions and hope to harness its insouciant majesty through sympathetic magic? Did they recognise the feral nature of humans lurking beneath the surface and find kinship with the lion? Did it just look cool? I don't know. But I do know that there was something of a cottage industry in Late Pleistocene Germany making images of cave lions out of mammoth ivory.* More realistic representations of cave lions are also known from southern Germany. These are smaller mammoth ivory sculptures almost designed to fit into the hand. Some really do show signs of rubbing and wear, as if they had been constantly stroked for reassurance like a good luck charm.

What does this all mean for our understanding of cave lions? Well, this über-successful species flourished in Britain until the end of the Pleistocene, when populations crashed nearly simultaneously in Eurasia and the Americas. Lions are a native British species, which is sort of mind-bending. They should be here but aren't.

We are seeing something similar happening now in Africa. Lions there have suffered an incredible population crash over the last century. There may now be fewer than 30,000 lions left in all of Africa. This is a reduction of 90 per cent in the blink of an eye. Unique subspecies have been entirely swept away. The North African Barbary lion, which lived along the Mediterranean coast, was last seen in the 1940s. It used to be said, when European and British railway lines were first connected in the nineteenth century, that 'a man who has dined on Monday in London can, if he likes, by making best use of express trains and quick steamers, put himself in a position to be dined on by a lion in Africa on the following Friday evening'. The Cape lion, last seen in the 1870s, was persecuted by Boers and Brits who alternately decried it as a pest and coveted it as a trophy. The Middle Eastern lion, last seen in the 1950s, was the

---

* As well as ivory mammoths, horses and even ducks ...

subject of the exquisitely carved Nineveh reliefs of Ashurbanipal,* king of the Assyrians, but was wiped out in its homeland.

Lions everywhere are edging towards endangered status. People are now talking about translocation projects to help manage declining populations. This may be the only way to restore lions to areas where they have gone extinct. The most endangered and isolated lion population, in the Gir Forest of India, is in dire straits. This tiny huddle of lions is the last remnant of a population that once stretched all the way from the Maghreb. The Indian government is determined to translocate some of the Gir lions to another site as an insurance policy against a catastrophic epidemic, climate change, poaching or any of the hundreds of issues that plague wild animals. Perhaps surprisingly, the Gujarati regional government is set against this, seeing the Gir lions as a focus of local pride. Why should other states that allowed lions to be hunted to extinction get some of their precious lions? The situation is at an impasse and there are still only a few hundred lions left in Gir.†

---

* The Nineveh reliefs can be found in the British Museum and are one of the wonders of the ancient world. Lion hunts were a royal prerogative for the Assyrians, and like the worst pampered playboys of our own time, they were recklessly indulged. The Nineveh carvings show how a canned hunt was set up, in the seventh century BC, with crated lions released to be leisurely shot or speared from chariots. The lions are beautifully rendered and full of pathos. One lioness crawls on her front paws, paralysed from the arrows severing her spine.

† The danger has accelerated, as recently two dozen Gir lions have been found dead from unknown disease, perhaps distemper caught from feral dogs. The clock is ticking.

Something similar, if more ambitious, has recently been put forward for the tiger. Molecular studies have shown that the extinct Caspian tiger that was found in a narrow corridor of habitat between Turkey and China until the 1980s is actually a close relative of the surviving Siberian tiger.* Some researchers now want to transplant Siberian tigers back to the territory vacated by Caspian tigers, with sites in Kazakhstan volunteered as a proving ground. So far, nothing concrete has happened.

I think it's obvious that no one is ever going to be on board with returning lions to Britain. But it might be interesting to think why this should be the case. We already live amongst lions. You're never more than 100 kilometres (62 miles) from a lion in Britain, thanks to the number of zoos and safari parks that hold lions legally. I prefer not to think about the number of illegally held lions there could be.† Rewilding lions in Europe (or North America) is a dead end because people rightly don't want to live with the risks that sharing space with a large hypercarnivore brings. We still have a lot to learn from those who do live with the risk. In some areas an

---

* Some people mistakenly think that white tigers and Siberian tigers are the same thing. Siberian tigers have the vibrant orange and black of normal tigers, as their homeland in eastern Siberia is only snowy for part of the year. White tigers are genetic freaks produced from constant inbreeding of captive tigers. One wild male white tiger was caught in India in the mid-twentieth century and the line was founded by backcrossing him with his own daughter to fix the allele. White tigers have no conservation value and often suffer from hideous deformities of the skull.

† It's been said that there are more tigers in illegal captivity in the state of Texas than there are tigers living in the wild in their natural range. The black market in big cats is huge.

uneasy truce has formed. But whenever there is conflict between humans and animals, it will always be the animals that have to give way. I can't see any solution to the erosion of wild lions in their natural range. Ecotourism and safaris may keep them safe for a little while. Fencing farmland and using specialist guard dogs may reduce the persecution for a little while. But piece by piece the battle will be lost until either humans or lions are gone.

CHAPTER FIVE

Woollies

Woolly mammoth
Extinct in Britain *c*.12,000 BC;
extinct globally *c*. 2,000 BC

Woolly rhino
Extinct in Britain *c*.12,000 BC;
extinct globally *c*. 12,000 BC

> Elephants are useful friends,
> Equipped with handles at both ends.
> They have a wrinkled moth-proof hide;
> Their teeth are upside down, outside.
> If you think the elephant preposterous,
> You've probably never seen a rhinosterous.
>
> Ogden Nash, 'The Elephant'

The hunter's mouth was dry. Every time she parted her lips to speak to the others, precious moisture wicked away into the steppe air. Nobody needed to hide, but everybody had to stay together. Following the well-worn mammoth trail to a watering hole, they could see a small herd a mile or so ahead. The matriarch, several older daughters and a sub-decade youngster. They were downwind.

Their people had performed this dance many times. Mammoths had been sent by the Great Mother to feed them, to clothe them. Providing dung to burn for heat, ivory to be made into weapons and jewellery. Because of the wisdom of the Great

Mother, the mammoths never ran but waited for them, knowing and embracing their fate. Only when the right words were spoken and the first spear was thrown would the mammoths trumpet their alarm. It was the song of the dance.

The dozen hunters walked purposefully towards their quarry, knowing that whichever direction the mammoths took, they could see them on the treeless plain. Before long they were close enough. The mammoths all eyed them with curiosity, the matriarch tossing her head and looking at them first with one eye and then the other, trunk writhing like a headless snake.

The lead hunter shouted the words. Drawing back her arm, two fingers locked the flimsy spear shaft to the atlatl. Flying quicker than the eye could follow, the spear whipped towards the youngest animal. The ivory-tipped shaft exploded through the shoulder blade and straight into the heart. Immediately after, the other hunters unloaded their spears of wood, horn and ivory into the rest of the herd. Some hit true, others only superficially buried themselves in the thick skin.

While the youngest animal was haemorrhaging, blood matting the black hair and making it glossy, the other mammoths lolloped away in confusion. They could be tracked later. One mammoth, whose back legs had buckled and collapsed, had been struck in the spine and forehead. Vainly it tried to pull the spears free with its trunk. The lead hunter, wrenching her spear from the shattered shoulder blade of the dead mammoth, ran towards the other paralysed animal. With both hands she thrust her weapon straight into the mammoth's face, at the base of the trunk. The enormous arteries that fed the blood-hungry proboscis severed, the mammoth gave up, drowning in its own blood.

Working fast to strip the skin and fillet the meat, each hunter knew their job. When everyone had filled their baskets, *the rest was left for the cave lions. They didn't have to heft the weight far: their camp was only a few miles back in the bend of the river, and one or two could go ahead to call for hands to help.*

The lead hunter had the glory of the mammoth's tongue. They cut it out to honour the hunt and so that the spirit of the mammoth could not speak ill of people in the afterlife. Such a muscular appendage was also a great delicacy. When the cave lions had taken their fill, they would come back to gather the bones – for marrow and for heat. Without trees, bones and dung were how people made fire here. Tracking and using the herds was how the Great Mother had made the world. She had given them the dance, and it would go on forever.

If I could somehow insert a graphic sex scene into the previous passage, it would be a passable sequel to Jean M. Auel's seminal *The Clan of the Cave Bear* series. Such 'palaeofiction' is an established genre, and even the late great Finnish palaeontologist Björn Kurtén dabbled.[*] I really think that scenes like the one I just described were commonplace, and part of the reason why humans were able to spread over the planet, especially in the northern mammoth steppe. Wherever we moved after leaving Africa, we found elephants: woolly mammoths and straight-tusked elephants in Europe, stegodonts in Asia, mammoths and mastodons in North America and gomphotheres in South America.

---

[*] His *Dance of the Tiger* is a good read, and has moderns, Neanderthals, mammoths and sabretooths, all the while using his vast knowledge of Pleistocene life to create a realistic vignette of the Ice Age.

They're all gone now, of course. Wiped out in the geological blink of an eye. It's a fundamental tenet of the overkill theory that humans were responsible, and I think the evidence is incontrovertible. All the proboscidean extinctions cluster around 14,000 calendar years ago – a global phenomenon from Portugal to Patagonia. I think the explanation lies in cultural change and, more importantly, in having a large enough human population that culture can be retained and transferred. We know that humans had the skills to hunt mammoths – there is abundant enough evidence of their success at this – but before humans reached a sweet spot of population density and connectivity, those skills could easily have been lost.

I think that Late Pleistocene people ran on what was basically a mammoth economy – following the herds wherever they went, taking enough that the population was eventually not able to replace itself. That economy would have been greased by the meat the mammoths provided, giving enough calories for people to not have to spend all their time hunting. It would also have been cemented by ivory, for trade and for use, to provide raw materials to make art and weapons, utensils and tools. The exquisite portable art of the Ice Age that we have inherited has usually been made from mammoth ivory: the *Löwenmensch*, the carved animals (including many mammoths), and the enigmatic Venus figurines.*

---

* Venus figurines are carved representations of women, usually either obese or pregnant, with wide hips and ample chests. They have been found all over Europe and into Siberia. The most famous example, the Venus of Willendorf (Austria), is made from limestone,

That's the light side of ivory. The dark side is that it is perfect for crafting lethal implements: lances, spears and daggers. We have those from the archaeological record too.

The bit of creative flow that starts this chapter was heavily influenced by data from two sites in particular. Yana RHS* is a 32,000-year-old accumulation of bone and stone in the high Arctic, on the Russian mainland south of the New Siberian Islands. Excavation here uncovered not only that people were able to survive within the Arctic Circle at the height of the Pleistocene glaciation but part of the reason why they were able to thrive. The Yana RHS cultural accumulation has some stone points and cutting implements (not made from flint, which is absent from the region, but from slate and quartz) and plenty of cut and scraped bones from mammoths and other megafauna. Just a few hundred yards from where the human presence is concentrated, a mass graveyard of mammoth bones is heaped. Mammoth graveyards are known in the high Arctic. Rivers like the Lena and the Indigirka washed mammoth bodies downstream for thousands of years and piled them up in dips and bends only to be discovered in modern times.

That's how the Yana RHS mammoths were discovered. Hardy ivory miners scour the landscape with the spring thaw, looking for tusks and bones to sell to a greedy market. They found the Yana RHS

but similarly shaped Venuses from Hohle Fels in Germany and Kostenki in Russia are made of mammoth ivory.
* Don't worry, this mysterious acronym will be explained later.

bones and reported them. Mammoth ivory is in huge demand, especially in China and Japan, to fill the gap left by a diminishing elephant population.* Bizarrely, mammoth ivory is entirely legal to use. I suppose that mammoths, being extinct, are therefore not in danger of extinction. It does lead to a thorny legal problem, though. Ivory carvers can claim that their wares are from mammoth tusks to avoid prosecution, even when they are using ivory hacked from the faces of poached elephants. Luckily, the biology of tusk growth has left a signature in the ivory itself. Schreger lines, first identified over two centuries ago, can be seen in cross-sections of tusks, leaving a beautiful mandala pattern of criss-crossing circles. Formed by the cells that laid down dentine in growing tusks, each species produces a different pattern, allowing them to be identified with near certainty. It's this quirk of biology that allows governments to regulate ivory.

Yana RHS, however, is different. It's not a natural accumulation but evidence of human lethality. The mammoth bones dumped at Yana RHS have plenty of cut marks, but what's more is that they have evidence of how multiple mammoths were killed. Two right scapulae (shoulder blades) from two different mammoths have stone and ivory points still implanted in the bone, evidence of projectiles hurled with

---

* Ironically, the very first written record of mammoth ivory comes from Chinese books of the fourth century BC. The availability of mammoth ivory for carving originally encouraged hunting of elephants for ivory, and now the rarity of elephants is pushing for the use of mammoth ivory again.

massive force into the living creatures. Another right scapula, from a younger mammoth, has an oval of bone punched right through it. Whatever weapon was used, it would have ended up penetrating the lungs and heart. The angle of entry for these spears seems to have been nearly perpendicular or coming in on a downward trajectory to the animal, suggesting that the hunters were using thrown weapons rather than thrusting weapons (which would have an upward trajectory). After successful hunts, the Yana RHS people must have taken the bones back to their camp at their leisure, stacking them like a woodpile to be used when needed. Fresh bones burn very well thanks to all the grease within them, and mammoth-bone fires were probably how these people stayed warm during the Arctic nights. The only strong evidence for mammoth consumption within the cultural layer, away from the main bone pile, is in the shape of five tiny, fragile pieces of hyoid bone. The hyoid is the bone found floating at the top of the neck that provides anchor points for the tongue. It must have arrived in the Yana RHS camp with mammoth tongue and then found its way into the midden there. Tongue is a very fatty meat, with lots of calories, which is prized today in many cultures. It's not a big leap to suggest that the tongue of a mammoth might have had some special cultural meaning to Pleistocene people – otherwise, how do you explain the relative abundance of tongue bones in the absence of any other small bones?

The other instructive site, Sopochnaya Karga, is on the edge of Yenisei Bay at the border of the Taymyr

Peninsula, and was home to a single young male mammoth from 48,000 years ago. The mammoth was nicknamed 'Zhenya' after one of the discoverers and is phenomenally complete. As is often the case, initial superficial studies suggested that it had died of natural causes.* Further investigation gave a different story. As with Yana RHS, 2,000 kilometres (1,250 miles) and 16,000 years away, there was massive trauma to the shoulder blade, this time the left scapula. Zhenya had been hit at least three times. As with Yana RHS, there was puncturing of the ribs too. Strangest of all was the skull. Zhenya was not a healthy mammoth – his teeth were deformed and he only had one tusk – but there was also evidence of human damage here too. His cheekbones had been pierced by a sharp instrument that had left a perfect cast in the damaged bone. The angle of the hole means it must have been caused by a human thrusting downwards from above onto a prostrate animal. Exactly like someone trying to deliver a *coup de grâce* on a fatally injured mammoth. Parts of Zhenya's jawbone had been smashed away too, likely by humans trying to access the tongue. His unitusk hadn't been taken away entirely, but the hunters had used it to try and make ivory flakes. By knapping at the tusk *in situ*, they could generate cutting tools to fillet the meat – pretty economical.

I think at this time the far north was running entirely on mammoths – fuelling a 'Bone Age' technology

---

* Given variously as death from fighting another young bull mammoth, or, as male elephants usually leave the herd about the age Zhenya died, the stress of this event giving him a heart attack!

reliant on ivory and bone rather than stone. There are plenty of other sites with reams of mammoth bones that show how important every bit of mammoth was. And that's only what we find preserved. There's nothing to show the potential use of mammoth hair to make ropes and snares, skin to make clothes and yurts, or fat to fuel lamps.

Those mammoth graveyards I mentioned earlier would have been plundered too, as they are now. When you are living on a knife-edge, no resource should be wasted. At around the same time as Sopochnaya Karga and Yana RHS, people were using mammoth bones the way later people would use bricks, building comfortable living spaces out of skulls and tusks.* Neat stacks of jawbones with vaulted tusk columns in skull sockets are a regular feature of the Late Pleistocene archaeology of Ukraine, Czech Republic and the Russian plain. Inside, some huts contain what could be instruments made of mammoth bones. Resonant skulls and dense shoulder blades daubed with chevrons of ochre were used as drums and beaters to make music. Everything could be made of mammoth.

I admit it's depressing, hearing about all the uses dead mammoths were put to. Living mammoths must have been incredible, jaw-dropping, awe-inspiring creatures. With elephants as a guide, it's fair to assume that mammoths must have had complex societies with

---

* The mammoth-bone huts are probably made from bones scavenged from natural graveyards. The thinking behind this is that the dating of different bones gives ages many hundreds of years apart. Explainable if they are from a natural death accumulation; slightly more difficult to explain if they were all hunted.

advanced communication skills. Elephants can recognise human speech from friendly and unfriendly groups based on their experience with them and know whether to avoid or ignore. They can recognise themselves in mirrors, something only primates and a few other species can do. Heartbreakingly, and unlike any other species except humans, they seem to recognise death. Elephants that find elephant bones (but not those of other species) will caress them with their trunks, pick them up and investigate them. It's probably best not to spend too much time thinking how mammoths would react to seeing the bone and ivory huts of Pleistocene humans. I don't think they would have liked them.

Undoubtedly, living mammoths would have been a big preoccupation for humans living alongside them. It's probably because of this that mammoths are one of the most common components of cave art.* The representations are so realistic that aspects of their soft tissue, only recently confirmed from frozen mummies, are apparent. The mammoth trunk is often painted as different from that of African and Asian elephants. African ellies have two similar-sized finger-like extensions at the end of the trunk that they can use for dainty pickups. Asian ellies have just one finger. Woolly

---

* There is one possible representation of a mammoth in the very sparse record of Pleistocene British art. If you close your eyes and squint and think happy thoughts, Gough's Cave in Cheddar Gorge, Somerset, could have a mammoth. It's a suggestive (but not very convincing) carving that, like a lot of Pleistocene art, looks to have used the natural lines of the cave wall to fill in the gaps. That, or pareidolia.

mammoths are shown in the art of Rouffignac and Chauvet in France with two fingertips but with the upper finger much bigger than the lower finger. How this difference would have affected their dexterity we'll never know. Mammoth trunks preserved in mummies also have a unique adaptation to the dry steppe environment. Endearingly nicknamed the 'fur mitten', the trunk has an expansion of skin something akin to the hood of a cobra, near the tip. Since those delicate trunk fingers would have been in contact with the cold ground, getting food all day every day, the 'fur mitten' might have been an adaptation to help it keep warm when the trunk was coiled.

For similar reasons, mammoths' ears were small (the exact opposite of African elephants' enormous lugs), as a way of conserving heat. African elephants can push blood through their giant ears, like radiators, to get rid of excess heat. Mammoths needed small ears to stop this happening.

One very necessary adaptation that mammoths had which doesn't get a lot of attention in the literature, but is known from mummies and from cave art, is the anal flap. The intense, dry cold of the Pleistocene was merciless in its ability to strip mucus membranes – any mucus membrane – of moisture. For an animal that processed as much forage and produced as much dung as the mammoth, this meant that even loosening of the bowels could lead to a life-or-death loss of water. Evolution's solution: the anal flap. A thick plug of skin that covered the anus like the lid of a cafetière, providing minimum air exposure to the sensitive mucosa and stopping moisture loss to the outside.

The shaggy hair and humped back in all the cave images make mammoths look like walking carpets, or banthas.* They are also adaptations to the mammoth steppe. The neck was full of insulating fat, giving mammoths their dowager's hump. They were also very furry. Dense underfur and long guard hairs made them look much bigger in life than the bones would suggest and helped to trap body warmth.

For a species so synonymous with being woolly, it seems kinda weird that the earliest mammoth bones actually come from Africa, but there you are ... Mammoths and elephants all come from Africa. One branch, the African elephants, have stayed there. The Asian elephants and mammoths left.

Britain, luckily, has a fantastic and deep mammoth fossil record. West Runton in Norfolk has one of the world's best-preserved examples of the woolly mammoth's direct ancestor, the steppe mammoth (*Mammuthus trogontherii*). An amazingly complete skeleton, about 700,000 years old, it was painstakingly excavated in the mid-1990s from the beach at Runton by a crack team of mammoth experts. The old male would have stood about 4 metres (13 feet) at the shoulder and had probably died of natural causes after a long and fruitful life. Although, the fact that one of his knees had dislocated and become infected a few months before death probably hastened his end. Our old friend the cave hyaena (see Chapter 2) had

---

* In *Star Wars*, the bantha on Tatooine that gets ridden by a Tusken Raider was actually just an Asian elephant in a furry costume. Apologies if you aren't a *Star Wars* fan and none of the words in that sentence made a jot of sense.

Map 4: *UK and Irish sites where mammoth bones have been found, along with calibrated radiocarbon dates where they are known.*

availed itself of a free meal and left an overabundance of bony coprolites littered around the bones. There are even signs that other steppe mammoths had moved the bones around, as elephants do today, mourning his loss. Today, West Runton is a beautiful stony beach. I've been there myself and stared in awe at the cliffs that gave us a mammoth.

The bones can tell us a lot about mammoths and their evolution. DNA can complement that story. Unfortunately, steppe mammoths like the one from West Runton are too old to be routinely used in genetic studies, but their descendants, the woolly mammoth (*Mammuthus primigenius*) and Columbian mammoth (*Mammuthus columbi*), are just right. The millions of bones still left in the permafrost have meant that mammoths are the go-to taxon for ancient DNA studies. The first Ice Age DNA ever recovered came from woolly mammoth bones way back in 1994, and since then mammoths have been at the vanguard of such studies. DNA has given an insight into mammoth life that bones could never match. For a long time, mammoths have been reconstructed with reddish-ginger hair, based on preserved fur from Siberia. This wasn't the case when the mammoth was alive. The orange hair colour is just the final breakdown product of the molecules that made up the pigment in the hair. Think of it like the white marble statues of antiquity. When they were first made they were painted in gaudy colours but the passage of time has scrubbed any sense of vitality, leaving just the white stone behind. It's the same with mammoth hair: it wasn't orange. Groups in Germany have identified one mammoth gene known to influence hair and skin colour, a gene called *MC1R*. In life it codes for a receptor that switches on the production of different kinds of melanin pigment. In mammoths it appears to have come in two forms: a wild-type and a mutant. The predominant form would have in all likelihood given mammoths dark chestnut-brown or black hair.

The mutant form would have given some mammoths a platinum-blonde look.*

Fishing out functional parts of the genome from mammoths isn't easy, but because there are so many bones, they have been used as a proof of concept study many times. Mammoth haemoglobin, the molecule that carries oxygen in the blood, has been reconstructed from mammoth DNA. Even more impressive than that, it has been *resurrected* in the laboratory. A Canadian group used bacteria to express the mammoth haemoglobin protein and study its properties. Although haemoglobin's job is just to carry oxygen round the body, it does this in a wonderfully dynamic way, more like an active enzyme than a passive transporter. It has evolved to work cooperatively so that if one molecule of oxygen binds, the entire shape of the molecule changes, which makes it even stickier to further oxygen molecules. Cooperative binding like this ensures that haemoglobin can scoop up all the oxygen it can carry in the lungs and deliver the oxygen where it is needed in the tissues. In living creatures, haemoglobin also has to deal with differences in temperature that affect how sticky it is to oxygen.

It turns out that mammoth haemoglobin behaves very differently to elephant haemoglobin. At low temperatures, the mammoth molecule was much better able to release its oxygen payload than the elephant versions. As you can imagine, a mutation like this would be very

*The same gene has been looked at in Neanderthals and confirmed that some would have had light skin and ginger hair. The Neanderthal study came out in 2007 and since then every Neanderthal reconstruction I've seen has had white skin and red hair.

advantageous to a mammoth living in the cold north of the Ice Age. It is just one of the many evolutionary adaptations that the living animal would have possessed to make life slightly easier. And the only reason we know about it is because some Canadian scientists brought a tiny bit of mammoth protein back to life in a test tube.[*]

An alternative approach using small bits of DNA from many different mammoths has given us an insight into how their populations changed through time. As genetic diversity is a function of population size, one can be inferred from the other. In mammoths, the size of their population mostly stayed constant, as far back as we can tell. Generally, mammoths did well throughout the whole of the Pleistocene. No obvious declines that could track the change in global temperatures are apparent. There is just no signal. Mammoth populations proceed along an even keel until all of a sudden, they don't. Exactly the kind of signal you would expect with a sudden, catastrophic extinction that had everything to do with people.

There are, however, two populations that prove exceptions to this general picture. St Paul is a small island (100 square kilometres or 40 square miles) in the middle of the Bering Sea. When Beringia was a great steppe grassland, it would have been a towering high point on the plain. Rare mammoth bones and teeth have been found here and dated too. The youngest comes from a

---

[*] Another gene, *TRPV3*, involved with temperature sensation, has also been Lazarused up in the lab. Unsurprisingly, it showed increased tolerance to cold temperatures: mammoths literally couldn't feel the cold.

mammoth that was alive around 4,500 BC, nearly 7,000 years later than the last mammoths on the continent, as far as we can tell. Humans never visited St Paul until modern times, so are off the hook in this case. St Paul is just too small. Likely the mammoths were bunched up into an ever-smaller parcel of land as the sea level rose at the beginning of the Holocene. Eventually there was just not enough food, not enough fresh water, and not enough space for the mammoths to survive. We just missed them.

The other place is Wrangel Island. It's an enormous island of 7,500 square kilometres (2,900 square miles) in the Arctic Ocean. This is where the very last, very lonely mammoths lived and breathed their last. They were still living here in 2,500 BC. And then they weren't. I talked to my colleague Dr Patrícia Pečnerová about what Wrangel is like. Patrícia has worked on the mammoths of Wrangel and visited the island too. Her fondest memories are of just how beautiful the place is. Compared to the icy deserts and lunar scenery of Greenland and Ellesmere Island, she describes Wrangel as 'like a garden'. In the summer it is flushed with colourful flowers, with musk oxen grazing and Arctic foxes gambolling. Talking to her made me want to visit. Although Wrangel also has one of the highest concentrations of polar bears on the planet, so you have to be careful. Patrícia told me that on Wrangel, mammoth bones are everywhere. She would travel around on an amphibious Argo with her driver, spotting bones and tusks just lying on the ground, waiting to be plucked up.

We have a genome of one of the very last mammoths of Wrangel, thanks to some of Patrícia's colleagues.

Unlike its mainland relatives, this mammoth shows very clear signs of a bottleneck at the Pleistocene/Holocene boundary, right about the time that sea levels were rising in Beringia and cutting Wrangel (and St Paul) off from the rest of the world. There may only have been one mammoth family on the whole island, outlasting the rest of their species. Unsurprisingly, this extremely limited gene pool meant that the Wrangel mammoths were very inbred compared to mainland mammoths. But that doesn't mean they were in danger of dying out just because of genetics. Yes, they'd lost about a fifth of their genetic diversity, but they were still more diverse than modern-day polar bears or tigers. The Wrangel mammoths were about as diverse as humans are today. In fact, the Wrangel mammoths are about as perfect a model as you could hope for when looking at the effects of extreme genetic bottlenecks on large mammals. In the interest of testing whether disease could have taken the last mammoths (as predicted by the hyperdisease theory), Patrícia and her team looked at immune system genes[*]

[*] The team looked particularly at major histocompatibility complex (MHC) genes, which are super-fascinating. In living mammals, MHC genes are expressed on immune system cells where they bind with fragments of viruses and bacteria that have been digested by the body. Bits of the broken viruses and bacteria get presented at the cell membrane on top of the MHC. Here they can be recognised by T-cells, and antibodies effective against them can be produced. Because of their need to bind super-diverse bits of protein, MHC genes tend to be very diverse and vary massively from individual to individual. In humans, the MHC determines whether organs can be safely transplanted. If a donated organ has a different MHC to the recipient, then the immune system thinks it is just a giant bacterium and will actively break it down, hence tissue rejection.

in Pleistocene mainland mammoths and Holocene Wrangel mammoths. If disease really had a role in the Pleistocene extinctions, then the genes involved with recognising bacteria and viral proteins should show signs of strong selection. Unsurprisingly, the Wrangel mammoths, while lacking in diversity generally, didn't show any signs that there had been selection for variation in immune system genes. Disease can't have been a big concern on Wrangel.

That's not to say that the Wrangel mammoths didn't have issues. As a small population on an island a fraction of the size of their natural range, they accumulated mutations that could have led to problems. Think of it like a sinking ship, trying to jettison all the non-essential items to stay afloat, throwing off enough to reach a new equilibrium. For the Wrangel mammoths this meant that genes coding for smell receptors and pheromone proteins usually excreted in urine were the first to go. With such a limited mating pool, the usual methods mammoths (and elephants) use to assess a mate (smelling their scent-marked wee) became non-essential. It got chucked. At the same time, mutations that had no net positive or negative effect were racking up. In the Wrangel mammoths this meant a few of them at least had a mutation that conferred a beautiful silky coat.*

We just have to imagine what that would have looked like, rippling in the Arctic wind. But we don't

---

* The gene *FOXQ1* was jiggered. In mice this gives a silky, satin look to the fur since there is no pigment in the outer part of the hair. The Wrangel mammoths would have looked like they came straight from a shampoo commercial.

have to imagine staring at a mammoth eye to eye. Mammoths are the symbol of the Pleistocene, and I think it's purely down to the fact that you can see mummies of them, big and small, that speaks to people in a way that bones don't. To be perfectly honest, you can put bones together any which way, and I'm sure some people look at the work of anatomists with a jaundiced eye and no small degree of suspicion. With a mummy, everything is viscerally there. You can see the pores in the skin, the wrinkles, the nails. It's a meeting, not a viewing.

Mammoth mummies have an afterlife of their own. They get their own names, travel enormous distances and become stars in their own right. Some of them border on having a personality.

Lyuba: the celebrity. The most complete baby mammoth ever found. She went on a round-the-world tour and was fêted in museums in London, Sydney and Chicago. Discovered in the Yamal Peninsula of Russia by a Nenets reindeer herder named Yuri Khudi and named after his wife, Lyuba is gorgeous. Her only flaws are a shortened tail and a missing right ear, which were nibbled off by hungry dogs after she was found.* Poor Lyuba died young. She was only a month old when she floundered at the edge of a lake and asphyxiated on

---

* Stories of eating mammoth meat are legendary. And more remarkable than this, they are sometimes true. Most mammoth mummies get scavenged when first exposed. Native groups are alleged to have used mammoth meat specifically in their fox traps, as it was irresistible to foxes.

mud.* Nonetheless, she has told us a lot about mammoth life. Her passing also left the kind of clues that could easily have been spun into a sinister conspiracy theory. She died in spring, but her gut had plant remains that are found in autumn. The most likely scenario? Coprophagy – the act of eating excrement. It's a normal behaviour in baby elephants who usually eat their mother's dung as a way of getting the intestinal bacteria they need to properly digest grass. Adult mammoths probably did it too, when access to fresh graze was restricted. If Lyuba's mother had eaten old mammoth dung, and Lyuba had eaten her mother's dung, then that would explain how autumn plants got into a spring baby.

Dima: the poster boy. If you've seen any pictures of a preserved baby mammoth, then it was probably Dima. In black and white, casually posed, lying on one side in the mud where he was found in 1977. Before Lyuba, Dima was the best mammoth body around, subject to pioneering studies to learn about this Ice Age baby. Like Lyuba, he was very young, perhaps only eight months old, when he slipped into some melted permafrost, drowned and was preserved.

Yuka: the fighter. Yuka is unusual in that she was around a decade old at the time of her death, whereas most mammoth mummies were either very young newborns or

* Something similar may have happened to a group of British woolly mammoths excavated from Condover in Shropshire. The adult and juvenile remains seem to have come from a kettle hole – that is, a small lake formed from parts of a melting glacier. The Condover woolly mammoths are some of the last mammoths known from Europe, dating to just over 14,000 years ago.

adults. Yuka has most of her hair and is spectacularly fluffy. Russian scientists cut out her brain (why not?) and found it to be incredibly well preserved, down to the level of individual neurons. Cause of death is unknown, but on her shoulder, throat, back and hind legs she has deep parallel claw marks that could come from a cave lion or conceivably a sabretooth.

There's also Buttercup and Khroma, Masha and Sasha, Effie and Adams, and more are being discovered every year as the permafrost melts.* Mammoth mummies grab all the attention, but they weren't the first extinct species to be brought to Western attention from the ice-soils of Yakutia. The other woolly, the woolly rhinoceros (*Coelodonta antiquitatis*), was actually discovered first. And it was only by a series of lucky coincidences that it made it out of Siberia at all.

In 1772 a young German biologist in the employ of Imperial Russia, Peter Simon Pallas, was trying to plough the same furrow that Georg Steller had so successfully turned before. Exploring new realms and naming everything after himself, Pallas† had a passionate interest

---

* Most mammoth mummies and most mammoth bones from the permafrost are male. They outnumber females two to one. Patrícia and her colleagues found this out by checking the DNA of all their mammoth samples. It's likely because mammoths lived in matriarchal groups that drove out males when they reached adulthood. Outside of the protection of the group they would be more likely to get into trouble and end up preserved. Basically, boys are stupid.

† Pallas's cat is a star of memes. Pallas's cormorant was native to Bering Island and eaten into extinction by the same groups that obliterated the Steller's sea cow. Steller thought they were 'delicious' and ate them alongside sea cows on his enforced stay on Bering Island.

Above: A speculative drawing of Steller and shipwrecked assistants measuring and dissecting a dead sea cow on Bering Island.

Left and below: Madagascan beaches are still covered in many large chunks of elephant bird eggshell (left). Cave hyaena coprolites (*album graecum*) from Aurignac Cave, France (below).

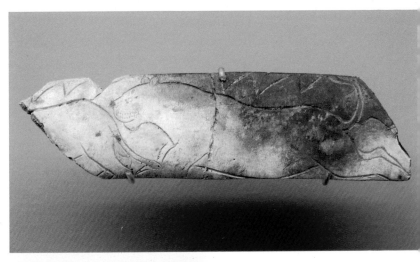

Above: A pride of running cave lions has been beautifully carved onto a piece of rib from La Vache Cave, France. Note the tufted tails and realistic expression.

Left and below: The enigmatic *Löwenmensch*: Pleistocene humanity's greatest work of art (left). The controversial *Homotherium latidens* canine from Creswell Crags. Note the lethal serrations (below).

Above: King Ashurbanipal's lion hunt relief from the palace at Nineveh (approximately 645 you can almost feel the lioness's pain and determination as she drags her paralysed legs.

Below: Spartak, the mummified cave lion cub, was found in the Yakutian permafrost, perfectly preserved down to his tiny whiskers.

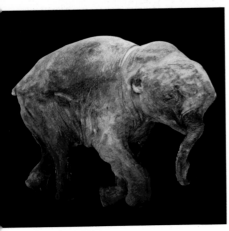

Top: A mammoth tusk pokes out of the ground on Wrangel Island, the last refuge of the mammoths. Finding these megamammals isn't usually this easy!

Above and right: Lyuba was just one month old when she died, 42,000 years ago (above). A mummified woolly rhino foot from the Siberian permafrost (right).

Above: A truly magnificent skeleton of a male shelk with full antlers. Observe the huge vertebral processes needed for muscles to hold the neck up.

Below: Blue Babe in his display case at the University of Alaska, showing the approximate pose in which he was found in the permafrost.

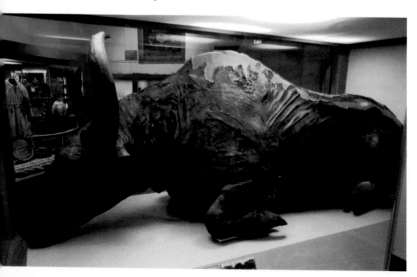

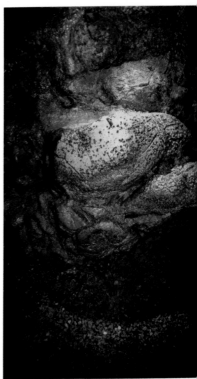

Top: The base of the Gundestrup cauldron depicts a male aurochs with a huntress and dogs. The missing horns may have been ivory.

Right: *Bärenschliffe* ('bear-shine') in Charlottenhöhle Cave, Germany, caused by millennia of rubbing by hibernating cave bears.

Below: Cave bear footprints and a hibernation nest preserved in the dust of Chauvet Cave, France. It looks like they've just woken up and walked away!

Top: This mosaic from Roman Yorkshire depicts a she-wolf suckling Romulus and Remus. I don't think the mosaicist had ever seen a wolf!

Left: A beautiful northern lynx in the Bavarian Forest, Germany. Lynxes' coat colour varies from silver to orange or brown.

Below: The Ardross wolf photographed soon after discovery. Such bold lines show that the artist was familiar with the wolf's form.

Above: A man holds beaver 'testicles' in his hand while another beaver attempts to castrate itself in this medieval bestiary image (left). Lord Howe Island horned tortoises had armoured tail clubs and seem to have gone extinct on this small island before humans arrived (right).

Below: Proud Heck cattle and Konik ponies graze in the Oostvaardersplassen, the Netherlands.

in extinction (unlike Steller, who was content just to cause it). While passing through Irkutsk, at the southern end of Lake Baikal, he was shown the remarkable head and feet of an unknown beast that had been discovered. Some Yakutian hunters had found the enormous corpse on a tributary of the Lena River far to the north. Unable to take the whole thing, they had cut off the head and feet and presented them to the local magistrate. The local magistrate, recognising their potential importance, had forwarded them on to government authorities in Yakutsk. Yakutsk kept a foot for themselves, and the head and other feet were then sailed 2,000 kilometres (1,200 miles) down the Lena to Irkutsk and the governor, arriving a few weeks before Pallas. Luckily, Pallas had published extensively on mammals and could see straight away that the head was that of a rhinoceros. It was exquisitely preserved, if a bit whiffy after its long journey – eyelids, abundant hair, tongue. It was the first Pleistocene mummy seen by a trained Western observer. Pallas had to send it back to the Academy *sehr schnell*. As it had already started to go off, he decided to dry it in an oven as a prerequisite for the journey to Moscow. Unfortunately, the head and feet were so well preserved that the ancient fat within them melted and, like a bad barbecue, one of the feet was charred beyond recognition. It was the first time in 14,000 years that woolly rhino had been cooked by people. The head and remaining two feet made it to the West.

That was the first mummy of a woolly rhinoceros. Despite their bones being found at many sites together, woolly rhino mummies are much rarer than mammoth mummies. They've averaged only around one a century since 1772. The second mummy was found in 1877 on a tributary of the Yana River. In a bizarre mirror of the first, only the head and legs were kept and shipped to Irkutsk. Its new home, Irkutsk's museum, completely burnt down in 1879, and the head only survived because a few weeks before the fire it had been on tour as part of an exhibition in Moscow.

The third and fourth rhino mummies aren't even from the permafrost. Starunia is a small village in Ukraine, with mines of ozocerite or 'earth wax', a waxy mineral that used to be a source of paraffin before oil. In 1907 and 1929 two complete woolly rhino mummies were found down the mine, not frozen but pickled. The mine had formed from an Ice Age swamp with ozocerite and brine seeps that had acted as a natural trap for large mammals. After they died in the swamp, the salty brine preserved the flesh and a coating of ozocerite sealed them perfectly. In contrast to the permafrost mummies, the Starunia rhinos are naked as a baby. The salt and wax have dissolved the hair and horns but left the skin and internal organs unscathed. There is a cast of the larger female rhino, as she was first found, above the Exhibition Road entrance to the Natural History Museum in London. It always makes me feel slightly sad when I see her. She is on her back with her legs kicking up in the air, ears sticking out at an alarming angle. It looks like she is having a bad dream.

Woolly rhinos and the Natural History Museum are indelibly entwined in my mind for other reasons too. When I was a fresh PhD student, one of my first assignments was to work on the best-preserved woolly rhino ever found in Britain. The Whitemoor Haye

Quarry rhino had been scooped out of the earth by a very surprised JCB operator in 2002 and passed on to the collections at the Natural History Museum. My supervisor wanted a sample to check for DNA, and I was the wide-eyed myrmidon taken along for the ride. Anxious about setting foot in the hallowed space of the museum', which had been my spiritual home since I'd visited it as a small child, I remember being a nervous wreck. Nothing prepared me for the experience of going behind the scenes. After meeting the redoubtable curator of fossil mammals in the public part of the building, we were taken on a trip that I can only compare to visiting Narnia or Hogwarts. In the museum, behind the giant *Megatherium americanum* skeleton that leans casually on a tree trunk in the hall of ichthyosaurs, there is a small, unobtrusive door, out of bounds to the public. It leads to the secret part of the museum behind the scenes. Thanks to our friendly curator, we were ushered into a fossil wonderland filled with countless rows of enormous rolling stacks, crammed full with bones from prehistory that held unfathomable secrets. Here, the skin of an extinct ground sloth from Patagonia; there, the serrated teeth of a sabretooth cat; here, the skull of a *Toxodon platensis* that had travelled with Darwin on the *Beagle*; there, the skull of Piltdown man. Wonderful things! The rhino we had come for was down an impossible labyrinth of corridors in a lab space full of workbenches and fume cupboards. I remember distinctly setting up in the fume cupboard with our small drill, using it to excise a chunk of bone from one of the long bones. One of the things I didn't expect was the smell. Drilling into well-preserved bones usually

Map 5: UK sites where woolly rhino bones have been found, along with calibrated radiocarbon dates where they are known.

gives off a pungent stink of burning hair. The proteins preserved in the bone are heated up by the spinning drill bit and burn acridly. The pressure was immense to get it right, and I've never yet got over my strong dislike of cutting into ancient bones to get samples. It seems too much like vandalism to destroy something so precious and old. I still don't like doing it.

After all that *bùrach*, I never even got any DNA from the Whitemoor woolly rhino. I don't know why it didn't work. Others have done it much more successfully than I. Their DNA showed that woolly rhinos sit in the family tree next to the critically endangered Sumatran rhino.*

The twenty-first century has given up a veritable crash of rhinos. And for the first time, a complete permafrost rhino – not just the head and legs – came out of the ground in June 2007. The Kolyma woolly rhino mummy was unearthed by gold miners on a tributary of the Kolyma River, north of the Arctic Circle. She doesn't have a name, but the Kolyma female is superb. Even freeze-dried she weighs more than one tonne, and in life she would have been nearer to two tonnes in weight.

The princess of woolly rhinos, and the only rhino mummy to get her own name, is Sasha. Sasha is a peach. Only around a year old at death, the left-hand side of her body is well preserved while the right is absent. She has the most gorgeous strawberry-blonde curls: thick and lustrous. Square-mouthed as the white rhino, she has been taxidermied into a lifelike pose. I say lifelike but she is faintly comical. As if Damien Hirst had chanced upon a mutant Highland cow and bisected it. Even her horns are cute – stubby nubs on the end of her snout. They've

---

* Confusingly, the Sumatran rhino isn't just found in Sumatra but also in Borneo and peninsular Malaysia. Or was. It is probably extinct in peninsular Malaysia. An optimistic estimate is 80 left in the wild over their entire range.

been worn on the top from Sasha continually rubbing against her mother while nursing.

Adult woolly rhinos have enormous, sweeping front horns 1 metre (3 feet) long, and equally impressive if significantly smaller secondary horns. Just like with modern rhinos, these weapons are made of flexible keratin, like hair and nails. But woolly rhino front horns are flattened and sword-like in outline, not conical. Conveniently, they have alternating annual dark and light bands, enabling anyone to calculate the age of the owner by simply counting. Using this developmental quirk and examining preserved horns, it looks like woolly rhinos may have lived up to 30 years, which is a lot less than other species of rhino around today.

The horns themselves have been found isolated and alone in the permafrost for a long time. Pallas saw some before he ever encountered his rhino head. Native Siberians (and later some Europeans) associated them with legends of giant Siberian birds of prey, the horns simply evidence of their enormous claws. Apparently, whatever their source, they were deemed useful by Yukaghirs and used to make hunting bows, the elasticity of the resulting product supposedly producing a superior weapon.

Rhino horn as a commodity has a long pedigree. Remember Yana RHS, the mammoth dump? Well, RHS stands for Rhino Horn Site, because that's what first encouraged excavators to dig. The discoverer, Mikhail Dashtzeren, picked up a magnificently worked foreshaft of rhino horn, honed into a lovely cylinder with bevelled ends. It's about 50 centimetres (20 inches) long and would have been just the end part of a much longer weapon. Like with the Clovis culture in North America, composite, multipart sophisticated weapons were the semi-automatics of the Yana Pleistocene. At the front would have been the pointy end made from bone, ivory or stone. This would have been hafted and glued to a short foreshaft.* The foreshaft would then be hafted to a main spear, maybe 2 metres (6.5 feet) long. The idea was that the foreshaft, the business end, could be swapped out – reloaded, if you will. A modular weapon with each part having a different purpose could be fixed, like the ship of Theseus, many times over.

The use of rhino horn by Pleistocene people raises an interesting question. Given how reluctant a woolly rhino would have been to give up its horn, how did they get them? I'm sure scavenging is a possibility, but I'm certain that woolly rhinos would have been actively hunted too. Just like mammoths were. And for all the same reasons: meat, fur, bone and raw materials.

However, in comparison to mammoths, there seems to be very little evidence of woolly rhino hunting. It is there, though. Wulanmulun in Inner Mongolia is a Late Pleistocene site that seems to be a bone dump or food midden. It's spectacularly dominated by bones of woolly rhino, smashed to pieces and with lots of cut marks on them. I think if horn was coveted, like ivory, then

---

* Made of rhino horn at Yana, but there were also similar-sized examples made from mammoth ivory at Yana. Wood and antler are known from other sites. Another rhino horn foreshaft has recently been found several hundred miles away on Bolshoy Lyakhovsky Island. The horn was only 16,000 years old (i.e. 16,000 years younger than Yana RHS), which means people were using rhino horn for a hell of a long time.

Pleistocene people would have been as capable of taking down a rumbustious rhino as a miffed mammoth. Even more likely if the rhino was solitary and the mammoths were in a herd.

The poor rhinos, with their producing only one or two young[*] every few years, would have needed only the lightest pressure to slip into oblivion.

---

[*] The Kolyma rhino mummy clearly has just two nipples, so two is the maximum number of calves a woolly rhino could support.

CHAPTER SIX

Irish Elk

Extinct in Britain *c.*10,500 BC;
extinct globally *c.* 5,700 BC

> They've taken the skeleton
> Of the Great Irish Elk
> Out of the peat, set it up,
> An astounding crate full of air.

Seamus Heaney, 'Bogland'

It can be difficult to wish back the toothy and terrifying from extinction. There would be few who would hanker after the return of the cave hyaena to these shores, or to resurrect the cave lion and sabretooth. These animals had their own beauty, of course, and their cascading influence amongst the ecosystems of the Ice Age were essential to their functioning. However, some of the species we lost had a jaw-dropping and innocent elegance that is painful to think upon. A majesty that speaks from their naked bones. The cruelty of extinction is that as soon as it happens, we re-evaluate the worth of the species. Few people loved the Tasmanian tiger until it was utterly destroyed. The Portuguese sailors who devastated Mauritius cared little for the plight of the dodo. It's only with hindsight that the obvious qualities of these lost species come into focus. It's all too human to suddenly want what we can't have.

It's the same with the Irish elk (*Megaloceros giganteus*). I doubt there has ever been a more majestic sight in the natural world than a male Irish elk with antlers held high. This, the largest species of deer ever to evolve, had an antler span of nearly 4 metres (13 feet), held on a muscular and powerful neck. If there were a Platonic ideal for deer, the Irish elk would be it. Every other deer is lesser in comparison. Antlers with a broad, solid centre giving space to numerous tines coming off front and back that resemble nothing so much as giant's hands projecting 2 metres (6.5 feet) either side of the skull. A Venus flytrap sculpted in bone. In reconstructions it looks like the king of deer, with the noble countenance of the red deer rather than the goofy bulbousness of the moose. It has occasionally been seen in popular culture. For those who managed to stay awake through the recent Hobbit films, King Thranduil rides on what is clearly an Irish elk. A stuffed Irish elk can also be very briefly glimpsed in the trophy room of Charles Muntz in Pixar's *Up*.

To briefly touch on the mammoth in the room, the Irish elk is not exclusively Irish, nor is it an elk.* Like the raccoon dog,† maned wolf‡ and giraffe seahorse,§ the species is badly named. The Irish elk has been found

---

* In British English an elk typically refers to the moose (*Alces alces*), the current largest living deer species. In American English, the elk (or wapiti) is their subspecies (or species) of red deer (*Cervus elaphus*). Taxonomy is confusing! Thank Linnaeus that we have binomials to separate scientific names from common names.

† Not a raccoon or a dog but a kind of Asian fox.

‡ Not a wolf but a South American canid distantly related to foxes.

§ Not a giraffe or a horse. You get the idea.

abundantly in Ireland, and it was first thought to be an elk (i.e. a moose), but we now know it was found all across Eurasia and is nothing like a moose. In fact, several eminent authors have given a lot of thought to how to neaten up this obvious discrepancy. My favourite solution is that given by Finnish palaeontologist Björn Kurtén. With wonderful brevity he sliced the cumbersome Irish elk into the 'shelk'. With some judicious cuts he coined a neologism that suggested something impressive and deer-like but removed the incongruous and misleading nationality. I love it, and will endeavour to use it here from now on.

If, however, you do happen upon a skeleton of this species in a museum, the chances are extremely high that it came from Ireland. Ireland is where it was first identified and where circumstances have conspired to produce the vast majority of bones. People have been finding the enormous antlers since at least the sixteenth century (according to surviving records) and were surely digging them up even earlier than this.

In Ireland the vast numbers of shelk remains have come from underneath peat bogs. For centuries, peat, a vast mat of decaying mosses, has been a staple fuel for people and needed to be cut from the ground. At the bottom of many Irish peat bogs is a fine marl deposit – evidence of prehistoric lakes filled with sediment ground from the rocks by glaciers. In this sticky blue clay, bones of the shelk are found – hundreds of them over the past two centuries, stained black by the leaching of tannins from the peat above. Those cutting the peat came across all sorts of evidence of the great antiquity of this

landscape: bog bodies from the Iron Age,* medieval
butter† and Ice Age bones. I know of one tale where a
peat-cutter reported finding giants' bones in the clay to
the local worthies. Finding his story honest, if not
credible, further investigation showed them to be bones
of the giant deer.

Scientific study of *Megaloceros* began back in the
seventeenth century. Sir Thomas Molyneux, an Irish
doctor and natural philosopher, was the first to read a
paper about them to the Royal Society of London for its
Philosophical Transactions journal. The title of his paper is a
classic of the sesquipedalian style of the time: 'A discourse
concerning the large horns frequently found under
ground in Ireland concluding from them that the great
American deer, call'd a moose, was formerly common in
that island: with remarks on some other things natural to
that country'. Molyneux was an astute observer with
imperfect data, and came to the entirely reasonable
opinion that shelk were the same as the American moose,
an animal not well known at the time. Interestingly,
Molyneux knew that *Megaloceros* was not the same as the
Scandinavian moose: he had inspected a Swedish rack and

*Tollund Man, eulogised by Seamus Heaney, was found in a Danish
bog, sacrificed for unknown pagan reasons back in the Iron Age.
Many bodies have also been found in Irish bogs, including Old
Croghan Man, a veritable giant (he was 2 metres or 6.5 feet tall!)
with manicured fingernails and mutilated nipples.
† 'Bog butter' is a mysterious substance found in bogs in Ireland and
Scotland. It looks like it was deliberately stored in wooden
containers, perhaps as an offering to gods or just as a way to
preserve it. After millennia of burial, it still has a noticeably buttery
texture and smell. Some brave souls have even tasted it.

discussed the clear differences in size and shape. If he had known, as we now do, that the American moose is the same species as the European moose, I wonder what he would have thought. Extinction was not a recognised idea then. The idea of a benevolent Christian God allowing any species to 'be lost entirely out of the world since it was first created'* was, if not heretical, at least very naughty.

Molyneux was writing at a very turbulent time, with the previous century having seen civil war in England and colonisation and conflict in Ulster. Perhaps as a result of this, many tried to foster Anglo–Irish goodwill by sending tributes of awe-inspiring shelk skulls and antlers across the Irish Sea. Examples were sent to Kings William III and Charles II. Charles was so impressed that he had them installed in his 'horn gallery'† at Hampton Court, where according to Molyneux they 'so vastly exceed the largest of them, that the rest appear[ed] to lose much of their curiosity in being viewed in company with this'. The political importance of shelk remains perhaps explains why *Megaloceros* is found on the coat of arms of Northern Ireland, appearing sinister (on the left), with a natty collar tight round its neck. Read into that what you wish.

Not just gifts for kings, commercial exploitation of shelk antlers has continued to the present day. A quick glance online shows that *Megaloceros* material is still readily available for sale at a range of prices. In 2006 a fine set of complete antlers sold for £57,360 at

---

* These are Molyneux's own words from his paper when discussing his belief that the shelk was likely to still be found alive in America.
† Pretty much a trophy room.

Christie's[*] – no doubt destined for the modern-day horn gallery of some nouveau riche arse. There was such a lucrative market back in the nineteenth century that Irish dealers brokered shelk bones for huge sums of money to museums, collectors and those with cash to spare. Back then, with peat being cut daily by hand, the chance of finding shelk remains without damaging them was high. Specialist collectors even probed the marl with long iron poles, hoping to strike bone. Nowadays, mechanised peat cutting has stopped the ready supply. Still, in their heyday, every rich landowner in Ireland wanted a set. To show their mastery over the land. To flaunt their historical connection to the island, especially where it was tenuous. Such was the trade both nationally and internationally, that shelk bones still pop up in the most unlikely places. I've seen shelk antlers and skeletons in museums big and small throughout the country, in the biology department of Durham University and in a small cafe at a wildlife park (two sets!).

This profligacy really owes itself to the astounding numbers of skeletons dug out of the Irish marl. Lough Gur in County Limerick is said to have produced so many shelk bones that they were shipped to Liverpool and London to be ground into manure for their phosphate. This is, sadly, not the only example of sacrilegious destruction. In 1815, to celebrate the Battle of Waterloo, shelk bones were piled into a bonfire and burnt in County Antrim.[†] They could do this because, like many other Pleistocene bones, they still contained a

---

[*] Indiana Jones's voice: 'IT BELONGS IN A MUSEUM!'
[†] History is always repeating itself, as some Swedish band once said.

Map 6: UK and Irish sites mentioned in this chapter, along with calibrated radiocarbon dates for shelk bones from sites where they are known.

lot of organic material. So much, in fact, that according to legend, one clerical joker was able to extract adipocere* from a *Megaloceros* bone and make it into a soup that was served to the Royal Dublin Society. I can't find any record of whether they enjoyed it or not.

---

* A waxy, greasy substance formed by the initial decomposition of fat.

People were reckless with the bones, common enough that they had no value. The prize was the skull; especially that of the male with mature antlers. It's not possible to do justice in words to how impressive the skulls are and how coveted they were. Having said that, before their worth was recognised, some were put to *very alternative uses*. Nineteenth-century writers reported that one large rack (remember that shelk antlers could span nearly 4 metres or 13 feet) was used as a temporary bridge across a stream in County Tyrone. Other skulls in County Tipperary and in Newcastle, County Wicklow, were put into use as gates, keeping sheep and cows in their fields!

Suffice to say, the antlers were big and impressive. What the original owners used them for is one of those mysteries that has caused generations of biologists to cogitate deeply. And not so deeply. Conspicuous as they are, nineteenth-century natural philosophers assumed that the antlers had a lot to do with the extinction of *Megaloceros*. When the idea of total extinction was new and weird, a sense of blame was always placed on lost species for being so careless as to die out. Authors claimed that the antlers became so cumbersome that the deer could not lift their heads and, being unable to do so, went extinct from overhunting. So wide that they couldn't move through woodland and died stuck between trees. So heavy that they sank into the clay when trying to drink and formed the famous bog skeletons by their thousands. All total nonsense, of course.

These explanations were strongly tied to the idea of orthogenesis: a concept, now discredited, that evolution is linear with a specific goal. Here, the aim was bigger and bigger antlers and so the shelk was ultimately doomed anyway. But that's not how evolution works!

Selection pressure is so constant that all physical features are only kept as long as they actively help the possessor to procreate, or at least don't impede them from procreating. The antlers of the shelk were enormous, but they were advantageous to the owner, that much is clear.

Since orthogenesis is rubbish, what other ideas have been used to explain the giant antlers? Well, I think there are only really three possible explanations. Firstly, they might have no negative effect but be a neutral side effect of something that has a strong positive. This was the view of the famous evolutionary biologist Stephen Jay Gould. Gould was the first biologist to actually bother measuring the skeletons of shelk to see whether there was some explanation for the proportions preserved in the skeleton. He was looking for signs of allometry. Allometry is the idea that there are strong non-linear relationships between the growth of different body parts. That is, when you compare the difference in body size between a tabby cat, a lynx, a leopard and a lion, the brain size of all these cats increases according to the negative power of the weight of the animal: the bigger the animal, the smaller the brain proportionally. In terms of percentage weight, a tiny domestic cat's brain is a much greater proportion of the total than in a giant lion. Allometry is a big deal in deer. When Gould compared antler size to body size[*] across many members of the

---

[*] Gould used a number of measures of body size, including shoulder height (easy to measure in complete recent museum specimens) and the length of the radius. The radius is the bone in mammals that goes from the inner elbow to the wrist. It's a good proxy for height in living mammals but has the advantage that it can also be measured in fossil taxa.

deer family, he found a very constant relationship: antler length increases exponentially as size increases. What Gould put forward in his research is that, if this rule holds across all deer species, then perhaps it wasn't the size of the antlers that was important, but just the size of the body. Maybe shelk were under pressure to evolve a larger body size, and as an unintended consequence of this the antlers had to grow larger. Under this scenario, the antlers of the shelk didn't have any real purpose – they were just there because the animal's body was so gigantic.

A second explanation for the massive antlers is as simply giant billboards. In this case, they would be advertising how enormously attractive the owner was. It's not such a crazy idea. Fallow deer bucks swing their heads from side to side to show off their palmate antlers and intimidate rivals without ever having to butt heads. As any Madison Avenue executive could tell you, 'bigger is better', and for fallow and shelk, the largest antlers could be enough to see off any rivals. In this scenario, rutting becomes a non-contact sport and selection for the most impressive antlers would naturally follow. This explanation could also describe the weird architecture of the shelk antlers. In most deer, the skulls are orientated with the antlers growing over the back of the animal, and the male has to swing his head to show how impressive his antlers are. In shelk, they grow at almost 90 degrees to the skull, sticking out to the side. When the shelk is standing still, the enormous flat palms are exposed to their fullest extent with minimal energy expenditure.

The third explanation is perhaps the most obvious. They were used for fighting. Here, again, the architecture of the antlers themselves provides some possible clues. Because of the orientation of the palms and tines on shelk antlers, the head would have to be carried with nose pointing to the ground for them to lock with an opponent. Luckily, some intrepid museum curators have done the hard experimental work on this. Dr Andrew Kitchener is a shelk expert who works at the National Museum of Scotland, but back in the 1980s he was at the National Museum of Ireland. In the attic roof of this august institution, he and an assistant curator engaged in a very unusual duel. Each hefting a skull and antlers of *Megaloceros*, they attacked each other and tried to find a way of making the two opposing racks 'fit' in a biologically plausible way. With the face pointing down, the palms of two opposing stags come into connection, tines locking like hands in prayer. The primary brow tines then act to cover and protect the eyes. Applying force from the enormous neck muscles would allow pushing and fronting between the males, forcing one into submission. It would have been a clash of the titans. Bone striking bone heard for miles around, bellows and snorts echoing across the landscape from the extreme effort. I talked to Andrew about this, and he confirmed that playing with shelk skulls was extremely hard work. To save their arm muscles, the curators attached steel wires to the roof timbers and around the skulls, manoeuvring them like enormous marionettes.

Of course, fighting would have been a last resort. No animal wants to fight and invite potential injury; much better to establish dominance by symbolic display. Overall, I think all three explanations could have worked in synergy to explain the enigma of the shelk. Body size and antler size in close allometric relationship would have selected for large, antler-heavy males to defend their access to mating. Those giant billboards would have intimidated all but the most evenly matched males from engaging in combat, sparing most from risk of injury. As a last resort, two equal males would have crashed skulls in a momentous battle for access to females.

There is some additional evidence for this. Shelk ribs have been found with healed puncture wounds that could have resulted from being stabbed by an antler tine during combat. Study of accumulations of bones from bogs also shows that small males were more likely to have died from malnutrition. And malnutrition must have been a huge issue for these creatures. Like other deer, the antlers were shed and regrown annually. To grow 40 kilograms (88 pounds) of antler in a 150-day growing season, the males must have been putting down 60 grams (2 ounces) of phosphorus and 30 grams (1 ounce) of calcium per day! Per day! In comparison, a healthy adult human needs just 0.7 grams of calcium and 0.5 grams of phosphorus daily. Getting all this, every year, from their diet would have been a massive problem. Some of it could be resorbed from the skeleton, sure. A cross-section of shelk jawbones shows that they were almost circular and could have been a small repository for minerals. Everything else would have had to come from plant sources. Like with lions (see Chapter 4), stable isotopes have given us details on what the shelk liked to eat. It was a very picky eater, choosing the most nutritious browse and graze. It selected only the plants richest in essential phosphate and calcium. And it would have needed to.

Most of the shelk we have fossils of lived in Ireland during a period known as the Allerød interstadial. In the crazy ups and downs of the Pleistocene, the Allerød was a tiny, brief period of climate similar to our weather conditions today. Before the Allerød was a period known as the Older Dryas, and after the Allerød was the Younger Dryas. These two periods that the Allerød is sandwiched between were both extremely cold snaps, characterised in the pollen record of these times by the appearance of *Dryas octopetala* shrubs. These are cold-adapted plants that do well in tundra-like Arctic environments but are not very nutritious. Luckily, the Allerød was reasonably temperate and allowed for productive grasslands and open woodland to form. This kind of environment is not really found in the modern world but was ideal for the nutritional needs of plentiful shelk – warm, with lots of nutrient-rich grass to eat, but not too bosky. It's only really the Allerød interstadial that the Irish remains date to. It looks like it was a brief interlude that allowed the shelk to expand their population from some refuge in the south or east. Of course, at this time Ireland was still connected to Britain,* which was connected to mainland Europe. When conditions were favourable, shelk could have walked from their refuges in Iberia or Siberia all the way to County Mayo.

We know from ancient DNA (including from some of those Irish samples) that shelk and fallow deer share a common ancestor but split from each other perhaps as

---

* Probably. Whether the Irish Sea was ever low enough to expose a land bridge between Britain and Ireland is contentious. Certainly, sea levels were low enough for some species to cross (like bears, shelk, hyaenas and shrews). However, there are some strange omissions from the fauna of Ireland. Snakes are the obvious one – kept out by Ice Age geography, rather than St Patrick – but there are others. The tiny, inoffensive field vole and dormouse didn't make it across either.

early as 10 million years ago. Fallow deer are the only
deer with similar antler structure to shelk, so this makes
intuitive sense. As an example of how science works, it's
worthwhile noting that this is not a complete consensus.
A study from back in 2005 used DNA to show that
Megaloceros was a red deer! Ancient DNA is a difficult
and fastidious field of work. Apart from the difficulty of
finding specimens that have any DNA in them at all, the
number one problem that ancient DNA researchers have
is the issue of contamination. Ever since the first ancient
DNA studies were published, the spectre of contamination
has stood at the elbow of every ancient DNA worker.
Look at it like this: when you are trying to get DNA out
of a 20,000-year-old bone, you also need to worry about
the DNA in the air, the work surfaces, the chemicals and
test tubes you use and other samples in the laboratory.
The techniques we use to get very old DNA would
much rather work on DNA that is only a few days old
and has come from a skin flake shed by the person doing
the analyses! In something like a shelk bone, there may
only be a few dozen DNA molecules per gram. In a
human skin flake, there could be tens of thousands of
human DNA molecules per microgram. It's a massive
problem. Back in the salad days of ancient DNA work, it
was a slapdash approach to contamination issues that led
to studies claiming to have DNA from Cretaceous
dinosaur eggshell and fossil dinosaur bones.* We're a bit

* These studies came out in the early 1990s and were a direct
inspiration to Michael Crichton when he wrote his little-known
book *Jurassic Park*. Sadly, all the dino-DNA turned out to come
from the human researchers who did the work.

more circumspect these days, but occasional doozies get through. This happened with the red deer-*Megaloceros* paper. The lab where the work was being done was also generating huge volumes of modern deer DNA for use as comparisons. At some point, contamination occurred, and what the researchers thought was shelk DNA had actually been switched for red deer DNA. Remember this next time a hugely controversial DNA study is splashed across the news. Scientists are all too human and make mistakes like everyone else.

Although it is there in the fossil record, we don't really have a good idea of what *Megaloceros* was doing for most of the Pleistocene. First records date to about 400,000 years ago from sites in southern England and Germany. It appears pretty much in its final form. The earliest shelk are about as big as the last shelk; however, there are some populations that seem to have slightly different proportions in the legs and antlers. It's all complicated by the irritating fact that outside of Ireland, giant deer are rare in the fossil record. There's an embarrassment of riches from the Irish Allerød and then zilch from anywhere else. I think this is probably tied in to *Megaloceros* essentially being a rare animal when it was alive. In Britain it's easy to think of natural populations of herbivores as quickly achieving high density in the landscape. We've been fooled by the example of red deer in the denuded landscapes of the Highlands. The reds are always described as overabundant, destructive and in need of culling (usually by those willing to pay for the privilege). This is a highly artificial state of affairs that does not exist in 'natural' populations. When there are lynxes, wolves, bears and wolverines in the ecosystem, deer numbers are readily kept in check. It would have been the same for shelk in the Ice Age. As massive herbivores with a specific niche, they would have been uncommon at best. I think that, apart from during the mating season, you would have been lucky to see one.

Ice Age humans did see them, though. In the rich heritage of cave art that survives, there are painfully few examples of what could be *Megaloceros giganteus*. Two caves give us the bulk of our shelk images: Cougnac Cave in the Lot Valley,* and Chauvet Cave in the Ardèche, both in southern France. Despite the hundreds of images of bison and horses, and dozens of images of lions, mammoths and rhinos, this is all we have for the shelk. Still, they are magnificent. With a postmodern economy of line, and masterful incorporation of the curves of the living wall into the composition, the Pleistocene artists bring an extinct species to vivid life. In Cougnac, the scene has two males surrounding one female shelk. Bizarrely to modern eyes, the antlers are almost incidental to the composition. While the males do have the classic *Megaloceros* antlers – with the broad palms, the uniquely backward-pointing tines and the fringe of projecting fingers from the palm – they are not exaggerated in the way that modern artists might represent them. They appear minimised and small in conjunction with the bulk of the rest of the animal. However, the outlines of

---

* Not much from the Lot Valley has come to worldwide attention. Cougnac Cave is one; Jean-François Champollion is the other. The man who deciphered Egyptian hieroglyphics was born here in 1790, just a few miles from a cave that has so many enigmatic and undecipherable pictures. The cave was only rediscovered in the twentieth century, but I wonder what he would have made of it.

the three animals are similar: we can see the living shape of an extinct titan. The heads are held lower than in most skeletal reconstructions, with the head almost parallel to the main line of the back. Biomechanically, this makes sense given the heft of the skull and antlers. Breaking this simple contour is the most incredible hump. At the shoulders, an enormous slab of muscle and fat is depicted, almost as if the artist's mind had wandered and they'd accidentally drawn a small mountain. This hump is only hinted at in the tall vertebral processes of the skeleton, but we see it here, resurrected, from the mind of a human who had the privilege of witnessing this Olympian deer. Additionally, some of the colouration is preserved. Coat colour does not survive in the fossil record but it is recorded here. The enormous hump is shown in a dark colour, contrasting with the cave-coloured flanks. Thin strips of pigment radiate out from the hump in a pattern that is not found in any living mammal. Switching to the Chauvet images, the hump and colouration provide the key. The shelk in Chauvet have no antlers. Perhaps they are females, or males from late winter/early spring before the antlers begin to grow. But they do have the massive hump, and classic giant deer proportions. The Chauvet shelk also have a dark-coloured hump and radiating lines – a link to the Cougnac images and confirmation that this was a feature of the living deer and not a stylistic flourish. Humans saw them. They faithfully recorded them in their caves. They surely interacted with them occasionally as well.

I talked to world expert on giant deer Professor Adrian Lister of the Natural History Museum in London about *Megaloceros*. I've known Adrian for years, and he is the embodiment of every gentle, kindly professor stereotype you've ever read. Generous with his time, and with an encyclopaedia of information mentally stored and instantly accessible, he had lots to say about the shelk. I asked him why there is precious little of *Megaloceros* in cave art, whether there was evidence that we hunted them, and if Pleistocene people could have used those enormous antlers for anything. What he had to say was instructive. Adrian thinks that the rarity of shelk on the painted walls of European caves is due to its rarity in the landscape. This tallies well with what I said earlier about how the Irish shelk remains have misled us about its frequency in the biota. He takes a wider view, and told me that shelk are usually only found as isolated bones in the Pleistocene fossil sites of Europe and Siberia. I also pressed him on what physical evidence there was for human–shelk interaction. We have bones of cave lion and *Homotherium,* red deer and reindeer that have been modified by Pleistocene humans. What about shelk? Surprisingly, it seems that there is not a single cut mark, not a single butchered bone that can be ascribed to *Megaloceros* from anywhere in the Pleistocene. Adrian and other researchers have looked long and hard for evidence but found none. It seems frankly incredible to me that the humans who we know hunted mammoth, bison, aurochs and woolly rhino would have failed to make a meal of the shelk. Maybe there was some taboo covering its hunting. Maybe there just weren't enough of them for evidence to survive.

I was also intrigued by those prodigious antlers, the 'frontal furniture' as the ghastly Richard Owen once described them. Antlers were the plastic of the Pleistocene.

People used red deer and reindeer antlers for a huge variety of purposes. Knife handles, picks, atlatls, digging sticks, retouchers, etc. Since they are shed every year, antlers are basically a forageable 'fruit' that could be put to a variety of artificial purposes by people. For every male shelk living to a good age, you would expect a couple of dozen antlers to have been produced. Adrian has also looked into this and found no worked *Megaloceros* antlers in the Pleistocene record. He did give one caveat, though. Depending on which piece of the antler – be it palm, beam or tine – was being worked on, it would be incredibly difficult to assign it to *Megaloceros* over another deer species. Perhaps, he mused, there might be some small pieces of worked antler sitting in a museum drawer that could be assigned to species by looking at the DNA or proteins preserved within. It's not so crazy. Pretty much the exact same kind of work has been done on medieval combs made from antler. In the archaeology of Viking Age Orkney, there are Pictish-style combs and Norse-style combs made from antler. Some historians used to think that the Pictish combs could have been made from reindeer antler imported from Norway, providing evidence for peaceful cultural exchange during this time. Looking at the proteins in antler with mass spectrometry allows the differences between different deer species to be identified. Like DNA, each species has its own unique protein sequence. The proteins themselves were non-destructively removed by simply soaking the combs in a buffer. The proteins showed that all the Norse combs were made from reindeer, and all the Pictish combs from red deer. Each culture kept to its own. Perhaps sometime in the future, an interested young researcher could use a similar technique to look at what deer species Palaeolithic people were making their antler goods from.

Now, of course, those same shelk antlers are the ultimate finite resource. We're not going to be getting any more, because *Megaloceros* is extinct. In Britain, the very last shelk lived in Ireland and the Isle of Man at the end of the Allerød. It used to be thought that some survived a bit longer, into the Holocene (our current epoch), but this turned out to be due to errors in the radiocarbon dating of the samples. Until pretty recently it was thought that shelk were extinct globally by the end of the Allerød. This was blown out of the water by Adrian and some of his colleagues in a seminal study published in 2004. Newer, better radiocarbon dates showed that a population of shelk survived in the isolated Ural Mountains until 5,700 BC. Nearly 5,000 years after the last Irish shelk died a lonely death, its conspecifics eked out an existence in the remote mountains of the east. It's a puzzle. While the megafauna that had ruled the Pleistocene were dying out everywhere in Eurasia, this tiny population was still thriving. There is no good explanation for why. Analyses of the pollen from this area show that it seems to have kept a productive open woodland with grass and shrubs. Just the kind of habitat shelk liked, and which had disappeared nearly everywhere else. What special conditions allowed this microclimate to thrive in the Ural foothills are unknown. As far as we know, the last surviving shelk lived and died here, disappearing completely just before the start of the Neolithic period. The weird thing about this extinction is that it cannot easily be pinned on any of the standard explanations for the megafaunal extinctions. There was nothing crazy about the climate in 5,700 BC. Modern humans had been in the region for a long time. Perhaps the Ural foothills were just isolated enough that Mesolithic people couldn't be bothered to hunt *Megaloceros* there. British shelk disappeared with the drop in temperature from the Allerød to the Younger Dryas. Humans aren't really implicated. For one thing, they didn't arrive in Ireland until a couple of thousand years after the last shelk died. It's also strange because although British shelk disappeared, the almost equally large moose *Alces alces* thrived up until at least the Bronze Age.

We know that moose were hunted in Britain, thanks to several well-preserved skeletons. The latest of these is a preserved moose shoulder blade from the Mesolithic site of Star Carr, in North Yorkshire. This moose had been through tough times, thanks to us, and the bone showed evidence of piercing by a flint blade. What's more, the moose had survived this initial attack because the bone had tried to heal itself and plug the rude gap left by the spear. The second hunted moose is older (probably from the Allerød) and comes from High Furlong in Lancashire.[*] This individual is even more of a smoking gun: the spearpoints that were used to kill it were preserved alongside the skeleton. These carved bone speartips were lethal. One was found between the

---

[*] This animal has the dubious distinction of having a Wetherspoon pub named after it. The Poulton Elk is in Poulton-le-Fylde, less than a mile from where the moose was found. Inside the pub they have kitsch decor, including a papier mâché head that looks like a taxidermied red deer – a completely different species.

ribs, the other stuck in the leg. Death would have been pretty instant.

Why moose, with multiple evidence of hunting, survived later than shelk, with no evidence of hunting, suggests that perhaps humans were not directly responsible for its extinction. I'm not willing to let us off the hook so easily, though. Shelk were extinct in Ireland before humans turned up, but in Europe it's possible that shelk hunting did occur, though at levels too low for us to detect in the archaeological record.

Reindeer are another mystery. Oh yes, we had these weird deer in Britain too. They also lasted longer than shelk. Depending on who you speak to, they died out in the early Holocene (at the same time as moose) or vanished as recently as the twelfth century AD. That's a big difference (more than 5,000 years if you're counting). The only evidence for medieval survival of reindeer in Scotland comes from the *Orkneyinga Saga*, one of the masterpieces of Viking literature, which describes the lives and times of Orkney under Norse rule. Amongst all the sailing and slaughtering, there is a single sentence that describes the Vikings going to the Scottish mainland to hunt '*rauðdýri eða hreina*'. You don't have to be fluent in runes to translate *rauðdýri* as red deer and *hreina* as reindeer. In a very Clintonesque way, our understanding of when reindeer died out depends on what was meant by '*eða*'. Some translators give it as 'and', some as 'or'. Vikings went to Sutherland to hunt red deer and reindeer, or to hunt red deer or reindeer. In the first translation, we give the narrator the authority that they understand the difference between both of these species, and that both were available for the chase. In the second translation, the narrator disavows any authority and basically shrugs their shoulders in saying they were after some kind of deer, but who cares what kind? Very frustrating for the lexicographers, and the biologists. It's tantalising, but I don't think it matters particularly whether reindeer were lost in the Bronze Age or the Viking Age. Like the shelk and the moose, they aren't here now,* and we are all the poorer for it. What we have left is the homeopathic dilution of a shadow of the natural biodiversity, dominated by members of one particular family, the Bovidae, whom we meet in the next chapter.

---

* Actually, there is a herd of feral reindeer in the Cairngorms National Park that seem to be doing pretty well. There have been a surprising number of attempts at reintroduction in Britain (usually rich folks' estates) and hardly any of them have taken.

CHAPTER SEVEN
Bovids

Aurochs
Extinct in Britain first
millennium BC (Bronze Age);
extinct globally AD 1627

Wisent
Extinct in Britain during the Late Pleistocene

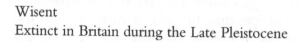

The cow is of the bovine ilk;
One end is moo, the other, milk.

Ogden Nash, 'The Cow'

Modern-day Britain seems very far removed from a
land of megafauna. The truth is that our largest wild
native carnivoran is probably the European badger and
our largest wild herbivore the red deer. However, if we
expand our definitions and look at things from a wider
point of view, the truth becomes slightly opaque. Cows
count as megafauna. Cattle (*Bos primigenius taurus*) have
not always been the gently mooing, friendly farmyard
denizens they are today. It's not generally talked about,
but like all domestic animals the humble heifer has
been modified from a wild relative. We know about
dogs and wolves, pigs and wild boar, but the case for
the cow is complicated by the fact that its wild relative
is now extinct. Cows come from something called the

aurochs* (*Bos primigenius*). I don't want to get all Gary
Larson but cows are generally underestimated by
people today. Don't be fooled. They kill more people
than sharks, bears and wolves combined. In Britain
alone, around five people are killed each year by cows.
They've been domesticated for more than 10,000 years
but tramplings and stampedes are still common enough
that you should be on your guard near Ermintrude. We,
as humans, always seem to underestimate the herbivores.
We're stupid like that. In modern Africa it's not the
lions, leopards or hyaenas that kill most people. By a
considerable margin of bodies, it's the Cape buffalo and
hippo that gore, mangle and maim. Aurochs were
another level entirely. A pretty good case can be made
that aurochs were the most deadly animal to live in
Britain, above even lions, hyaenas and sabretooths.
We know what aurochs were like because they were
described by literate, knowledgeable people and were
woven into the culture of medieval and earlier European
peoples. We know precisely how ferocious a wild
aurochs was because we have written sources that
tell us.

There are few more literate historical figures than
Julius Caesar. He encountered aurochs on one of his
bellicose jaunts into Germania. The man who bent
the known world to his rule and tamed the wild

* Aurochs can be used as singular and plural, but aurochsen is used
if the writer wants to be fancy. The name comes from the Germanic
for 'ur-oxen', or primal cow. Also the name of my Bobby Gillespie
tribute band.

Vercingetorix was terrified of what was, in essence, a wild cow. He wrote in his *Commentaries on the Gallic War*:

[Aurochs] are a little below the elephant in size,[*] and of the appearance, colour and shape of a bull. Their strength and speed are extraordinary; they spare neither man nor wild beast which they have espied.

They were strong and they were quick. One seventeenth-century account (recounted in Cis van Vuure's *Retracing the Aurochs*) says: 'They are so dazzlingly fast that after they have defecated, they even manage to playfully catch their manure with their horns before it reaches the ground.' Quite a party trick! These were also enormous animals. We know from skeletons that a full-grown male could easily look a 2-metre (6-foot) person in the eye. Their horns were absolutely gigantic: double-curved, like mammoth tusks, with spear-sharp tips pointing straight ahead. Deadly. Images drawn from life are abundant in the caves of Chauvet and Lascaux, and down through time to the much cruder woodcuts left in medieval bestiaries. We know from many sources that the bulls were black-coloured and the cows reddy-brown.[†]

People lived side by side with aurochs, and nowhere is this better expressed than in a stunning piece of European

---

[*] A slight exaggeration. We can forgive Caesar this little white lie. Aurochs were maybe twice as big as a modern cow. Still formidable but not exactly elephantine.

[†] This sexually dimorphic colour scheme is found in other living bovids, the Asian banteng and gaur.

Iron Age art, beaten out of solid silver. I'm talking about the Gundestrup cauldron, a priceless treasure from Denmark. Found by peat-cutters in a Jutland bog in 1891,[*] the Gundestrup cauldron is nothing less than a liturgy of pre-Christian beliefs, written in a symbolic language we can barely understand. There are deer-horned men holding snakes and torcs, crude elephants and lions, griffins, wheels, soldiers and cavalry. Some men are depicted playing the carnyx, the legendary war trumpet of the Iron Age. A giant dips a man head-first into a cauldron, holding him by his feet. A kneeling man wearing a horned helmet[†] grips a chariot wheel. It means something all right, but who knows what? It's definitely not from Denmark but was probably commissioned from expert silversmiths[‡] in central Europe and brought back to be used and repaired over several centuries. One of the repairs, to cover a hole in the bottom, is clearly from the hand of a different master metalworker. This circular plate was likely recycled from a horse's bridle and shows a downed aurochs. Legs splayed and twisted as it lies dying, the animal is lifelike and detailed. The head rises in relief out of the flat silver, breaking into the third dimension as it labours for a last

[*] Further proof that peat bogs are full of interesting things.
[†] The horns on the helmet, somewhat reminiscent of aurochs horns, look just like the ones Loki wears in the Marvel movies. Loki's helmet is based on real Danish examples of Bronze Age helmets from Viksø, north of Copenhagen. They were also found in a peat bog.
[‡] Although the bulk of the cauldron is silver, it was also originally gilded, though this has worn off. The individual plates were soldered to an iron ring to keep them together. The tin in the solder has been shown by isotopic studies to be Cornish.

breath. Around the dying animal, three dogs lie exhausted
alongside their sword-wielding mistress. Yes, mistress.
Despite the obvious breasts and long braided hair, some
who have studied the cauldron describe the aurochs
hunter as a man. It's not. The images are pecked into the
silver using a variety of tools, cross-hatched and stippled.
The bullock's bollocks are prominent. Horns and ears,
now missing, may have been made of ivory or glass. The
overall impression given by the work is of total exhaustion
after a struggle to the death. It tallies well with what
Caesar wrote centuries later. Elsewhere on the cauldron
is an external plate where three giant aurochs are about
to have their throats pierced by sword-carrying figures
while cats and dogs run together round about. Aurochs
were clearly an important part of the culture that
commissioned the Gundestrup cauldron. They remained
an important part of many European cultures right up to
their extermination.

The extinction process in the aurochs is a fascinating
and sad story. The last record of a living aurochs is from
AD 1627. This is more recent than the ascendance of the
Stuart kings, later than the Gunpowder Plot, younger
than the invention of logarithms. While they disappeared
from Britain before 1,000 BC, they held out much longer
in the wild forests and meadows of mainland Europe.
Caesar recorded the German aurochs, and other authors
reported aurochs from the French Pyrenees and Polish
woodland. The Polish animals were actually the very last
population. Beside the forest of Jaktorów in central
Poland, a rough-cut boulder now lies in commemoration
of the last wild aurochs. Jaktorów forest was a hunting
preserve, kept for Polish nobility from medieval times

and surrounded by peasant villages; it was a wild and dangerous place. The aurochs that lived here were free-roaming and untamed. However, since they were so rare and important, the Polish court were far-sighted enough to set up what was in essence conservation legislation to try and preserve them. Specially employed hunters supplemented their winter feed with hay and chased any aurochs that strayed into local farmland back to the forest. This controlled management of the last aurochs population was reasonably successful for a couple of centuries, providing enough aurochs to allow hunting by the king and his favoured few. The aurochs hunts were organised spectacles[*] with a pre-scripted finale – similar to the scene depicted on the Gundestrup cauldron millennia earlier. When the aurochs had been felled, but before finally expiring, a pantomime of death began. Local tradition had it that the curly hair found on the forehead between the horns had magical powers. This patch of skin would be cut from the dying animal and turned into belts that supposedly had the power to grant an easy labour to pregnant women. Only after the curls had been cut would the animal be killed. Immediately after death, the heart would be cut out.

Inside the heart of bovids and other ruminants lies a bone, the os cordis, which in life supports the aortic valves. This bone (or heart stone) from the aurochs also had the power to give pregnant women an easy birth and would be sent with the curls to the king. We know all this thanks to notes preserved by one of the only

---

[*] There is more than a passing similarity to modern-day 'canned hunts', given that the aurochs were habituated to humans.

learned men to visit Jaktorów and see aurochs alive.
Anton Schneeberger was a Swiss botanist who had
studied at the University of Basel. His letters also have
something to say about the legendary fierceness of the
aurochs. He wrote:

> An aurochs is not afraid of humans and will not flee when
> a human being comes near, it will hardly avoid him when
> he approaches it slowly. And if someone tries to scare it by
> screaming or throwing something, this will not scare it in
> the least, but while it stays in its place it will actually open
> its mouth, widen it and close it again quickly, as if it is
> making fun at the human for his attempt. When it is stand-
> ing in the road or somewhere else, one must go around it,
> even if one is driving a carriage, since it will not move off
> the road by itself.

Quite a contrast to Caesar's description. Whether this
was because the last aurochs were more used to humans,
or further exaggeration on Caesar's part, we just don't
know.

Despite the sustained management of the Jaktorów
aurochs during the fifteenth and sixteenth centuries, a
crisis was unfolding. Problems of local encroachment
on the forest, overgrazing, poaching, deforestation and
logging by local concerns were all having an extremely
negative effect on the wildlife.* The demise of the
aurochs plays out in a series of administrative inspection
reports. In 1566: 'The aurochs, which are in the forests of

---

* I hope this resonates. These are exactly the issues that we see in
the Amazon, Borneo, the Congo and many other biodiverse forests
today.

Wiskitki, need better living conditions; a shed for the winter needs to be built for them, but also, not so many herds of cattle should be permitted as is the case at present.' In 1597: 'We demand that the inhabitants of said village do not herd cattle and do not mow grass for their own use there, where the aurochs live [...] The [forest manager] has to ensure that our forests, where the aurochs live, are not destroyed by said subjects, so that the aurochs may retain their former habitat.' In 1602: 'In these forests at Jaktorów, aurochs are hiding, of which at this moment there are only four, which we have seen for ourselves, three bulls, one cow.' It is a classic example of an extinction vortex. Populations decrease in size, causing problems with inbreeding and magnifying even small losses of individuals to disease, old age or hunting, which then cause further population loss. In 1620 the very last aurochs bull died, and his horns were preserved as a monumental drinking horn.* After 1620 only a solitary cow was left. An 'endling',† she died of extreme old age in 1627. And so the aurochs was extinct.

Or was it? Domestic cattle are aurochs, albeit shaped and formed by people into a very different animal. Can we say that aurochs are extinct while cattle thrive? I think

---

* The drinking horn is in the Royal Armoury museum of Stockholm, taken as booty during the Swedish invasion of Poland (1655–1660). Many aurochs drinking horns are preserved. The largest known was actually found buried at Sutton Hoo, one of a batch of seven.

† Coined in a 1996 *Nature* letter by Rob Webster and Bruce Erickson. How sad that we have had to come up with a word to describe the last, lonely, doomed member of a species. And that the situation is common enough that a word is needed.

we can. I don't know of anyone who would argue that it would be fine for wolves to disappear as long as dogs stay around, or for wildcats to be exterminated as long as we keep the moggies. The legendary wildness of the aurochs, the characteristic that defined them as a living animal, has been lost. Wildness is a very nebulous trait. We know it when we see it. And cows don't have it.

We have many thousands of bones of aurochs. Tallying the numbers suggests that they were not terribly common during the Late Pleistocene and then mushroomed during the wetter and warmer conditions of the Holocene. Somewhat surprisingly, we have many more examples of hunted aurochs in the fossil record than for almost any other megafauna. Why should this be the case, when aurochs survived hunting pressures that caused the extinction of many other species much earlier? I think the answer is instructive. In western Europe, aurochs survived for 11,000 years more than shelk. That's 11,000 years in which evidence of shelk hunting could be lost to the elements and also 11,000 more years in which aurochs hunting could occur. The attrition of evidence of older hunts and the addition of newer hunts can explain the discrepancy and remind us that absence of evidence is not evidence of absence. Like the moose skeletons discussed in the previous chapter, there are aurochs skeletons with the weapons of their destruction still imbedded within their bones. A Mesolithic aurochs from Denmark[*] was particularly unlucky. This individual bull has two flint points stuck in two different ribs: one healed, the other raw, showing at least two different attacks were

[*] At Vig, a site near to Gundestrup but much older.

attempted. It also has a neat, circular hole blasted through
the shoulder blade, perhaps caused by a spear. However
terrifying aurochs were, people hunted them armed with
wood and stone. In England, the Neolithic site of Burwell
Fen in Cambridgeshire had a complete aurochs skull
with a flint hand-axe still stuck in the forehead. One can
picture the Stone Age hunter delivering the *coup de grâce*,

Map 7: UK sites mentioned in this chapter.

and then their intense annoyance at being unable to pull the axe back out.

A continent away in the famous Turkish site of Göbekli Tepe, more evidence has been found. This impressive ritual area is perhaps the first dedicated place of worship known in the archaeological record and is covered in a strange iconography of vultures, boars, scorpions and aurochs. One of the weird things about Göbekli Tepe is that the citizens who built it left the remains of monumental feasts littering the floor, either deliberately, for reasons unknown, or to cover the site in rubbish as some kind of closing ceremony. In amongst all the detritus there are dozens of aurochs bones, but one stands out. This humerus, the front upper leg bone, has a flint projectile stuck into the bone. This is strong evidence of human hunting, with the hunter aiming for the heart and lungs but hitting the foreleg instead.

The region around Göbekli Tepe is fundamental to the story of the aurochs for another reason. It's here that humans showed their mastery over nature's most formidable mooer by taming, and finally domesticating, the aurochs. The Near East, around Göbekli Tepe, is where cows were made. In probably only our second attempt at animal domestication (after the wolf/dog), hunter-gatherers decided that staying in one place and keeping cattle in a bounded enclosure was much easier than trying to hunt the buggers in the wild. It's hard to overstate how fundamentally the simple (to modern eyes) act of taming wild aurochs and keeping them nearby changed the course of human history. Without it, we might still be hunter-gatherers, moving from place to place with the seasons. The only reason people began to stay in one spot and create villages, towns and cities was to keep crops and animals. It is the starting point of modern civilisation. From that starting point with one or a few animals, probably 12,000 years ago, there are now around 1.5 billion cows. They are the most populous vertebrate species on the planet after chickens, and humans. You could say they've done pretty well out of domestication. From so simple a beginning, with a handful of aurochs, a global cow empire arose.

The story of cow domestication is fantastically interesting and has been teased out of bones and DNA. There were actually at least two distinct domestication events: one in the Near East, and one in India. The Indian domestication led to what are called zebu, indicine or humped cattle. They derive from the subspecies *Bos primigenius indicus*. Western cows come from *Bos primigenius taurus*, the Near Eastern subspecies. This difference is reflected in the genes. Because the subspecies that was domesticated (*taurus*) is different from the subspecies found naturally in western Europe (*Bos primigenius primigenius*), they are separate enough to show different mutations in their DNA.

Colleagues have actually extracted ancient DNA from aurochs and early domestic cattle and looked at what was going on. It's a complicated story that goes something like this. The first farmers expanded out of the Near East (the 'Fertile Crescent'), taking with them their crops and their new wonder-product: the cow.* These farmers had

---

* 'You can kill and eat it with little risk to life and limb! But wait! There's more! While the females are alive, they turn vegetation into milk that you can drink! Need leather to make clothes? Why not cow™? Tired of ploughing your land every year? Use cows™ and the work is done!

already been breeding cows/aurochs for a while and their animals were substantially smaller in size than the wild aurochs still around in Europe. The size difference is a good measure of whether bones come from wild or domestic cattle. The size of the bones also matches the genetics. The small cows have what is called the T-signal (or haplogroup, named T for *taurus*) in the genes they get from the maternal line (mitochondrial DNA), whereas the larger European aurochs have the P-signal (or haplogroup, named P for *primigenius*). Pretty clear-cut. This stark maternal signal seemed to show that all European domestic cattle came from that original Near Eastern domestication event. However, when techniques had advanced enough that we could look at genes from both parents (Y chromosomes and autosomes of the nuclear DNA), the story evolved. It looks like farmers started practising husbrandry once they had settled in western Europe with their Near Eastern cows. Some wild male aurochs were allowed to mate with the domestic cows and left signals in the ancient bones and in modern breeds. One cattle heritage breed, the Cabannina from Liguria in Italy, is a large and beautifully russet-black animal. In its genes it carries the P-haplogroup signal,* showing that its maternal ancestry can be traced back to European aurochs. An explanation for this could be independent domestication of Italian aurochs. Or it could be that the early farmers crossed

---

* It's a bit more complicated than this, actually. The Cabannina carry the Q-haplotype, which has probably come from native Italian aurochs, but no one has looked at fossil Italian aurochs to check ...

their cattle with Italian aurochs cows. DNA data from bones found in a Neolithic Swiss dwelling beside Lake Biel directly shows the act of this hybridisation. This tiny Swiss domestic cow also had the P-haplogroup signal showing that its direct maternal ancestor was a female aurochs. It makes sense to me. Those early farmers might not always have been able to get enough cows and bulls for their husbandry needs. When that happened they could take advantage of the wild (and free) aurochs that lived nearby.

Now we are in the era of sequencing complete genomes from extinct species, the aurochs was a natural target. Access to the nuclear genome – the part that has genes from both male and female lines – can tell us so much more about ancestry and evolutionary history. The first aurochs genome was completed in 2015, and the aurochs chosen for the task was British. The special bone came from Carsington Pasture Cave in Derbyshire, and is dated to before the introduction of farming to Britain, thus ensuring that the animal was purely aurochs with no contribution from early domestic animals. The data is illuminating. Comparing the genome of the aurochs to a range of modern European cattle breeds shows two things. Firstly, modern British heritage breeds like the Kerry, Dexter and Highland cow have a significant input from British aurochs, which is not found in mainland European breeds. Their distant ancestors must have been crossed with British aurochs. Secondly, compared to wild aurochs, domestic cows have been highly selected for genes associated with immunity, size, muscle growth and brain behaviour. During the last 12,000 years, this signal shows farmers have been breeding cattle to be healthier, bigger and beefier. Oh, and stupider too – all good

things for the farmers. This is also the reason researchers
are so interested in the genes of extinct aurochs. In times
past, pastoralists could cross their cows with wild aurochs
to introduce new genes into their stock – genes that
could provide useful traits for new cows. With the
extinction of the aurochs, this is no longer possible, and
the only pathway to aurochs genes is in the molecular
biology clean room.

This leads directly to one of the most oft-repeated
questions in the study of ancient DNA. Can we resurrect
the aurochs? The short answer is no. The long answer
is still no but with a few caveats. Gene-modifying
technologies are advancing quickly and scientists are now
in the position where they can mutate single base pairs of
DNA in living animals.* Because aurochs and cows are
the same species, I can see a future where farmers working
with a good understanding of cow genomics could screen
aurochs for interesting alleles (i.e. alternative forms of
genes) that enhance things like beef marbling, muscle
mass, leanness or other traits of interest to the farmer,
and then inserting the aurochs genes back into the
cows. Similarly, wildlife ecologists, archaeologists,
conservationists and romantics might be interested in
resurrecting a 'real' aurochs by modifying a cow genome
to express all the aurochs traits. This would be a long and
arduous process of identifying where cows and aurochs
differ in their genomes, then using technology to modify
the genes in an embryo, and bringing the embryo to

---

* For more about how exciting technologies like CRISPR–Cas9
can be used to 'de-extinct' lost species, read Beth Shapiro's excellent
How to Clone a Mammoth.

term. Would the resulting creature be anything more than an ersatz aurochs? It wouldn't have the behaviour of an aurochs, as there would be no aurochs to teach it how to behave. It would be a cow that looked like an aurochs. Would that be good enough? If it is, then surely a much easier path would be to simply use the traditional techniques of artificial selection to just breed a cow that looks like an aurochs. This has been done. And it's a pretty dark story.

Back at the beginning of the twentieth century, Ludwig Heck, the director of Berlin Zoo, had two sons: Lutz and Heinz. They grew up at one of the premier zoological institutes of the time, surrounded by animals and vets. Unsurprisingly, they decided to follow in their father's footsteps. Lutz continued at Berlin Zoo and Heinz moved to Hellabrunn Zoo in Munich. Both had become fascinated by stories of aurochs and separately decided to try and recreate the extinct form by cross-breeding different forms of cattle. Using cave art, written reports and woodcuts as a guide, the brothers set about mixing breeds to reproduce an animal that looked like an aurochs. To achieve this, Heinz mixed a number of European breeds together, including Highland cattle. Similarly, Lutz mixed Spanish and French fighting bulls with Corsican cattle. Their unsystematic approach focused on reproducing the coat colour, markings and horn shape of the original aurochs.

Unfortunately, as the brothers embarked on their projects, Germany was turning to the Nazi Party and their perverse ideology began to infect every aspect of German life. With this in the background, the two brothers began to follow very different paths. Lutz joined the SS and openly courted favour from Hitler's second in command Hermann Göring.* Obsessed with hunting and styling himself as *Reichsjägermeister* ('Reich Hunting Master'), Göring took a keen interest in the resurrection of the aurochs. He wanted to use the aurochs for political points, reintroducing it to the Polish forests where it had been exterminated, as a way of showing German mastery over their new territory. Using the legendary *Nibelungenlied* poem and its hero Siegfried as a guide, Göring set up his own wildlife park and wanted it stocked with all the animals hunted by Siegfried. He leaned on Lutz to provide elk, bison and aurochs to fill it. After he felt his crossing experiments had produced some success, Lutz reportedly boasted: 'The extinct aurochs has arisen again as German wild species in the Third Reich.'†

In contrast, Heinz did not go out of his way to appease the Nazis. He had been married to a Jewish woman during the First World War and was overheard saying he'd 'prefer one Jew over ten Nazis'. Furthermore, in startling contrast to his brother, he is said to have shot a lion cub given to him by Rudolf Hess rather than house it in Munich Zoo.

Nonetheless, both parallel breeding experiments were tainted by association with Nazism and did not really produce a resurrected aurochs as the brothers had hoped. The breed produced became known as 'Heck cattle', and was dark-coloured with curly hair on the forehead, a pale snout and a dorsal stripe – all characteristics that

---

* Lutz gave Göring a pair of lion cubs from Berlin Zoo as a gift.
† The 'Herd Reich', if you will. This amazing pun was employed by *The Sun* in their report of a failed attempt by a Devon farmer to import Heck cattle to his farm.

were probably found in aurochs. Heck cattle have been propagated and are still to be found in parks and collections today. They are all descended from Heinz's animals. Lutz's animals, like the Third Reich he embraced, did not survive the end of the Second World War but were all killed during an Allied bombing raid on Berlin.

At the same time, Europe's only other large bovid was barely hanging on. The European bison, or wisent (*Bison bonasus*), is the largest surviving remnant of the European megafauna. Currently Europe's heaviest wild animal, it can weigh over 800 kilograms (1,765 pounds). In the past, it was often confused with the aurochs but is an entirely separate species. Like the aurochs, one of the last populations was conserved in a Polish forest used as private hunting ground by the nobility. The forest of Białowieża had wisents until 1919. One other population, in the Caucasian mountains, made it until 1927. Most people think of bison (*Bison bison*, also wrongly called buffalo)* as an American species, emblematic of the push into the Wild, Wild, West. Nope. Europe had bison for millions of years before America. In fact, aptly, American bison are a very recent immigrant to the Americas, arriving only about 130,000 years ago. This is a legacy of what was once a truly cosmopolitan species, ancestor to both European and American bison – the steppe bison (*Bison priscus*). Now, nobody knows more about steppe bison than my friend Professor Beth Shapiro. Beth is without doubt one of the smartest people I've ever met.

---

* As the classic joke goes: Q. How do you tell the difference between a buffalo and a bison? A. Mainly through genome sequencing and comparative phylogenetic analysis.

She has a ferocious intellect and drive, which I've personally witnessed on many occasions. She worked in TV as a teen, received a MacArthur 'Genius' award, has been a National Geographic Explorer and was the first female Rhodes scholar from her home state of Georgia. Woe betide the naïf who underestimates her. I think I did once, but I got away with it. Beth and I had the same PhD supervisor, and she taught me more than almost anyone else how to work with ancient DNA and Pleistocene bones. Beth's thesis was on the genetics of bison. In her own words: 'Bison are the cockroaches of the Pleistocene.'

Steppe bison are probably one of the most common Ice Age fossils out there. And I can personally vouch for this. Many people think of the Ice Age as crammed with mammoths and mastodons, heaving with giant sloths and sabretooths. Actually, during the Pleistocene, vast swathes of Eurasia were more like the African Serengeti of today. Whereas in the Serengeti, most of the biomass is in wildebeest and zebras, most of the Ice Age Eurasian biomass was concentrated in bison and horses. During a field trip to Yukon in the early 2000s, I liaised with many gold miners and saw this for myself. The 'placer miners', as they are known, were a crucial link in our search for Ice Age bones. In frigid Yukon, gold is mined by searching for old riverbeds, where in the past gold dust and small nuggets had been washed out of the deposits and into the river. Over time, as with all rivers, these changed course and left the old riverbeds behind to be buried under millennia of organic detritus. Leaves, plants, soil and the rest. The miners identify old river courses and use enormous water cannons to blast away the frozen organic muck, or overburden, to get down to the river gravel that they then scoop up and process for the gold. While washing away the frozen dirt, something quite amazing happens. The run-off mud forms a channel and, as if by magic, bones appear at the sides, like precipitate from a solution. In the millennia that the organic material was accumulating, many bones and carcasses would have been incorporated into the mulch. Kept at frozen temperatures thanks to the Yukon climate, the bones are amazingly fresh-looking and perfectly preserved for DNA analyses.

I have spent many happy hours wandering amongst the muddy run-off, picking up countless bison bones, horse bones and occasionally other species. I was looking for lions and sabretooths but mostly just found bison. On cutting open the bones, some even had marrow still preserved, like stank jelly, on the inside. Thanks to the gold industry, Yukon has one of the richest collections of Pleistocene vertebrates in the world, including some amazing mummies (more on this later). Beth's work on bison, using many samples from Yukon, was a game-changer in our field. She used a combination of specimen dates (from radiocarbon) and DNA sequences to do something that no one had ever done before with vertebrates: track population change in a species through the Ice Age. Her work showed that bison had gone through a population explosion in the lead-up to the coldest parts of the Late Pleistocene, and had then declined in numbers on the approach to the Holocene. If you had asked a Pleistocene hunter-gatherer to place a bet on which of the megafauna they lived with looked like it was going to go extinct, it's possible they would have picked bison. Yet here we are, with bison as pretty much the largest species of megafauna found on two continents – a true survivor.

In amongst the Pleistocene muck discarded by the heirs of the Klondike Gold Rush, one very special specimen deserves more attention. Back in 1979 near Fairbanks, Alaska, a gold miner was startled to see animal feet sticking out of the mud he had been monitoring. He called in the regional palaeontologist to investigate further. That palaeontologist, R. Dale Guthrie, will forever be associated with the unique carcass he encountered that late summer day, which went on to consume a huge amount of his research life. 'Blue Babe'* was a nearly complete mummy of the steppe bison and one of the finest bodies to come to us from the Ice Age (see also Chapter 4). Guthrie's studies set the bar on what could be achieved from looking at a single animal. His intimate observations of Blue Babe uncovered an animal that, while superficially like its descendant bison in Europe and North America, actually had many distinct characteristics. The tail was much shorter (an adaptation to cold temperatures), the coat pattern was different and the horns were larger and darker. Subtleties of the head would have given Blue Babe a noticeably

---

* Named after the sidekick of legendary lumberjack Paul Bunyan in American–Canadian folklore: Blue Babe was Bunyan's faithful blue ox. 'Blue Babe', the steppe bison mummy, is also a beautiful purple-blue colour in places thanks to the coating of mineral vivianite. Vivianite sometimes forms on fossils as the reaction between iron in the soil and phosphate in the bone. I've seen it coat bones in many shades from electric blue to royal purple. It's shimmery and gorgeous.

different profile. I find that the most evocative of Guthrie's findings came from his detailed study of the circumstances leading up to Babe's death. While he was prepping the preserved skin, he found that a series of long parallel scratch marks had been scored into the skin of the rump and neck. Additionally, tooth puncture marks were found on the muzzle and back. The punctures occurred in pairs, about 8.5 centimetres (3.3 inches) apart. This was all the evidence needed to show that Blue Babe did not have a natural death. He was hunted. And the evidence pointed to the culprit. Alaska has many carnivores: red foxes, wolves, pumas, lynxes, wolverines and grizzly bears. Ice Age Alaska had more: cave lions, scimitar cats and giant short-faced bears. Only cats have claws sharp enough to have pierced bison skin. Only lions have canine teeth 8.5 centimetres apart that could match those found in Babe's hide. Blue Babe was brought down by one or more cave lions. Possibly two lions working together, one grabbing the rump, the other applying a classic strangling bite to the muzzle, exactly the way African lions take down Cape buffalo. Blue Babe was likely killed in early winter and the cave lions would have tried to gorge on the meat before the body froze. However, the quick freezing process was how the mummy retained so much tissue. Eventually, scavengers came in to try and get a free meal as the body became a bisonicle. There is also evidence that lions, perhaps even the same pride that brought Babe down, returned to try and glean a few scraps. While Guthrie was prepping the skin for further taxidermy he found a miniscule fragment of tooth enamel embedded in the flesh. The thickness of tooth enamel is a diagnostic character in carnivores. The chip of tooth was just over 1 millimetre thick. Too thick for anything except the carnassial* tooth of a cave lion. It had snapped off when the lion had tried to bite into frozen flesh and received a wicked toothache.

You can see Blue Babe at the University of Alaska to this day. One of the conditions that the mine placed on the scientists was that Babe must be displayed in perpetuity, and he is. Without this stipulation, it's possible that Babe would not have been taxidermied as part of the process of study. If Babe had ended up in a university freezer behind the scenes, we would all have been robbed of one of the greatest stories to come from the study of Pleistocene megafauna. Dale Guthrie tells the story in his book *Frozen Fauna of the Mammoth Steppe*. To celebrate the successful mounting of the blue bison and the completion of many years' study of the remains, a great party was held. Guests of honour were Björn Kurtén, Dale Guthrie and Eirik Granqvist the taxidermist. The crowning glory of the occasion was a rich stew served to the honoured attendees. The secret ingredient of the stew? A cut of meat from Blue Babe's neck, aged to perfection in the permafrost. For the first and last time since the end of the Ice Age, people dined on steppe bison. It apparently gave the meal a distinct whiff of melting overburden but was enjoyed nonetheless.

Blue Babe lived around 36,000 years ago, the heyday of the steppe bison, when they could be found from

---

* Carnassials are the technical name for the scissor-like molars and premolars found in some carnivores that act to cleanly slice through flesh.

England to Mexico. How the steppe bison slowly morphed into the wisent used to be a bit of a mystery. Still, some of the best clues were found in the art of the Pleistocene. Clearly discernible on the painted walls of France, Spain and Germany were two different forms of bison: the long-horned, beefy-looking steppe bison, and the weedier-looking, small-horned wisent. Were there two populations? Two European species? Two time periods? When we look at the painted roof of Altamira Cave in Spain with its hundreds of polychrome wisent–bison and compare it to the stark outlines of the Lascaux steppe bison, what is going on? It's only in the last few years that an answer has been suggested from the genetics of ancient bison fossils. People have known for a long time that something was weird in the genes of the wisent. Studies of the mitochondrial DNA, inherited down the maternal line, showed that the wisent had a form that looked similar to that found in cows, whereas American bison had a form that looked like that of the yak. What the fudge? The wisent and the American bison are clearly close relatives – just look at them! Why did their genetics not show a close match? It's a story told in the nuclear genes. Buckle in.

Nuclear genes* are those that determine what an animal looks like. We get them from both mother and father and they make up the chromosomes that sit in

---

* Don't be frightened of the word 'nuclear'. It just means 'kernel' in Latin as a useful way of describing the small centre inside things like cells and atoms. Know that when you go for an MRI (Magnetic Resonance Imaging) scan, this is just the sanitised application of a technology known in every other field as Nuclear Magnetic Resonance (NMR). It was changed so as not to scare people.

the centre of each cell in the nucleus. Sequencing of nuclear genomes from ancient wisent bones revealed a massive secret. What we know as the modern wisent is actually the hybrid offspring of a dalliance between steppe bison and the ancestor of the aurochs more than 120,000 years ago. This cross-species freakiness explains why the modern wisent has mitochondria that look like the aurochs's – at least some of the hybridising must have been between a male steppe bison and a female aurochs, who bequeathed her mitochondria to the entire wisent lineage.* But it gets even freakier. By painstakingly dating the bison bones from the Late Pleistocene and cross-matching them with both the genetics and dated bison images in art, the researchers found an amazing correlation. Before about 50,000 years ago and after 34,000 years ago, most of the bison were wisent-like. In the period between 50,000 and 34,000 years ago, the bison were mostly steppe-like. It looks as if the last steppe bison in Europe lingered on during the coldest part of the Ice Age before dying out completely (except in North America, where they became American bison) and were replaced by the wisent form, with its invigorating shot of aurochs genes that gave it the ability to cope with the different ecological conditions that prevailed as Europe warmed back up. Astoundingly, this ecological transition was recorded by early Europeans in their art! Testament to

---

* This makes total sense. In modern experiments, crosses between cows and bison produce infertile males and fertile females, so only the females have the potential to produce further hybrid offspring!

their incredible powers of observation, there is a clear difference between very early sites with the steppe bison form depicted, and the later sites with the wisent form. This leads to the inescapable conclusion that you might be able to give a ballpark figure for how old some European Pleistocene art is based just on what kind of bison is depicted. Isn't that mind-bending? It's also equal parts reassuring and depressing that despite the aurochs being extinct for 400 years, its genes survive both in modern cattle and European wisents. The aurochs is a genetic zombie, living after death in the DNA of cows and beefalos.

The fossil record tells us that the steppe bison went extinct in Britain sometime in the Late Pleistocene, and the wisent probably never made it over at all. However, one of the earliest bits of evidence for human presence in England comes from cut marks on a half a million-year-old bison bone from Happisburgh in Norfolk. Humans and bison must have rubbed along well enough as they survived until at least 400,000 years after this. A bison is also star of the only cave art ever found in Britain, from Pin Hole Cave in Creswell Crags. Along with a red deer and a few other scratches, these are the only examples of Pleistocene cave art known in Britain, so they must have been an important species over here for Palaeolithic people. Aside from the cool transition of bison forms recorded in cave art, they are one of the stars of one of the strangest scenes in the whole repertoire of cave art. Known as the 'bird-man' scene in Lascaux, it depicts a man lying on his back with a bird-like head and prominent penis, a duck with what looks like one very long leg, and an enraged wisent with exposed intestines and arrow-like projections sticking into its haunches and on the ground beside it. Close by, a serene woolly rhino defecates with tail raised high. Traditionally, it has been interpreted as a legendary hunting scene with some kind of mythic overtones. I like this explanation, though there are others. The image could tell the story of a hunter who took on a wisent but only managed to disembowel it. Understandably peeved, the wisent knocked the hunter down and killed him.[*] There is no good explanation for what the one-legged duck and free-flowing rhino have to do with hunting wisent. One of the great things about cave art is that anyone can be an expert. Parallels with modern hunter-gatherer customs don't necessarily tell us anything about the cultures that painted Lascaux. We'll never know for certain.

Amidst all the uncertainty, there are a few things we know. We know that bison, aurochs and humans shared a landscape in Europe for more than half a million years. We know that for most of that time bovids were hunted. We know that they survived for a while, while other comparable species didn't. The aurochs went extinct a couple of centuries ago; the wisent (and the American bison) would be extinct now if it weren't for immense efforts in conservation and captive breeding. We've seen the devastation we can wreak on wild species, but this

---

[*] Since being gored by a wild beast, although exciting, is not usually likely to cause tumescence, maybe this aspect of the image shows that the hunter was killed outright. Post-mortem bacterial fermentation in human bodies can cause the scrotum to swell spectacularly, leading to the appearance of erection.

information has only come to us thanks to the written records of bison and aurochs hunters. If we did not have those accounts, would we now be talking about their extinction due to the Little Ice Age? Or the Medieval Warm Period? Would we be trying to pin their loss to a climatic event like some do with other Pleistocene megafauna? I think we would.

CHAPTER EIGHT

Bears

Polar bear
Itinerant visitors during the
height of the Ice Age

Brown bear
Extinct in Britain by Roman times

Cave bear
Extinct in Britain during the Late Pleistocene;
extinct globally *c.* 22,000 BC

> Magnificent bears of the Sierra are worthy of their mag-
> nificent homes. They are not companions of men, but
> children of God, and His charity is broad enough for bears.
> They are the objects of His tender keeping.
>
> John Muir, Scottish–American naturalist

The great white bear of the north. *Ursus maritimus*.
Largest meat-eater on land. Never eats penguins.*
During the height of the Ice Age, polar bears would
have been occasional visitors to Britain. Using the vast

---

* Polar bears live in the Arctic and penguins live in the Antarctic, as
any fule kno. However, the original penguin was the great auk, a
flightless relative of the puffin and razorbill that lived in the North
Atlantic until it was hunted to extinction in the nineteenth century.
The first European explorers to encounter penguins in the southern
hemisphere called them penguins because they reminded them of
the great auk that they already knew.

expanse of sea ice that spread south to meet the glaciers on land, they could have walked from Svalbard to Stornoway during the Late Glacial Maximum. Some must have done exactly that, because we have evidence to prove it. In the far north-west of Scotland, in a place so remote that most Highlanders have never even visited it,* are the Bone Caves of Inchnadamph. Found in one of the loneliest parts of Sutherland, here there are few animals and even fewer people. The caves are the main draw, and they have provided a rich assortment of Pleistocene fossils. Unique in the Scottish fossil record, and perhaps in the British record too, is the only skull of a Pleistocene polar bear.† Found wedged under some rocks that fell from the roof of the cave, the characteristics of the skull tell us that it was from the white bear. Polar bears have very distinctive teeth; as the most carnivorous of the bear family, their molars have evolved to slice flesh rather than crush roots and so have a more blade-like aspect to them.

During the Pleistocene, when northern Europe was swathed in ice, polar bears could venture much further south. There are records of them in Denmark and even in Hamburg. It's not known if they were as reliant on seals and other marine mammals for food as they are now. What we do know is that in their expanded range they had a much greater likelihood of bumping into

---

* Well, I haven't.
† An ulna (arm bone) of an extremely large bear was found in Pleistocene gravel at Kew Bridge in London and initially placed as a new type of polar bear: *Ursus maritimus tyrannus*. Later reanalysis suggests it's actually an enormous brown bear. That's it for British polar bear fossils.

brown bears (*Ursus arctos*), their sister species. Today, you occasionally hear stories of 'grolar bears' or 'pizzly bears'. These oh-so-cute portmanteaus refer to the offspring of a mating event between the two species. Today, a warming climate is causing brown bears to venture ever further north, to places that would normally be covered in ice, where they encounter equally surprised polars. During the Ice Age, the natural expansion of the frozen north allowed polars to roam further south. What happened when the two species met has left signals in their genes.

It's been known for a long time that polar bears have weird genetics. Even back in the 1990s when researchers were looking at small snippets of mitochondrial DNA, something didn't make sense. In the mitochondrial DNA, the sequences from polar bears were more similar to brown bears from Alaskan islands than those Alaskan brown bears were to bears in Europe. That didn't make any sense. The polar bears should have been on their own distinct branch, but instead they were right in the middle of the brown bears. Only later, with advances in sequencing technology, did the story start to make sense. It seems that polar bears, like the wisent in the last chapter, are hybrids. All the polar bears around today can trace their ancestry back to an ancient grolar or pizzly sometime in the Pleistocene. The guilty secret is written in their nuclear genes. And it looks like the love nest for their trysts was Ireland. Only one other group of brown bears had a closer genetic relationship to polars than Alaskan brown bears and they were extinct Irish brown bears.

I was lucky enough to be involved with the study looking at British and Irish bears way back in 2004, while I was working on my PhD. Dr Ceiridwen Edwards, an expert on ancient bear genetics, who went on to write up the study, kindly invited me to do the replication work she needed. As mentioned earlier, one of the issues with working with any kind of ancient DNA is that it is incredibly easy to contaminate. DNA is in the air, on every surface, all over the place. When you are working with old bones that might only have a few molecules of DNA left in them, you have to be extremely careful not to introduce newer, better-preserved DNA from the environment. This is a huge problem when you are working with humans or domestic animals, as their DNA is absolutely everywhere. It's easier when you are working on bears and lions, whose DNA doesn't tend to be naturally in the surroundings. Still, there is always the potential for cross-contamination between your samples, and that's why controls, double-checks and replications are so vital.

Ancient DNA work takes this paranoia a step further and asks, 'Well, even if you get the same result multiple times, could it be something to do with the laboratory you are doing the work in?' External replication was brought in to control for this. Researchers getting interesting results could send samples to a different laboratory and ask them to try and generate the same results. It was for this reason that Ceiridwen asked me to drill holes in some very old brown bear vertebrae, which I gladly did, to try and reproduce the results she was getting. Which I did. Everyone was happy. Ceiridwen's Irish bears were super-interesting. They showed that along the maternal line, polar bears seemed to be an offshoot of the extinct brown bears of Ireland and

England. Further delving into the nuclear genes enhanced the picture. The mix of genes found in polar bears can be most parsimoniously explained by at least two hybridisations between brown bears and polar bears, with most of the gene flow going from the ancestor of modern browns into the ancestor of modern polars.[*] It's super-freaky. The idea that many of the species that we recognise today could be the result of hybridisation – like polar bears and wisents – was not even suspected until recently.

The catalyst has come from a better understanding of our own past. Ancient DNA from Neanderthals and the mysterious Denisovans[†] has shown that we humans are hybrids too. Like polar bears and wisents, every human with non-African ancestry can trace a little bit of their DNA back to a different species. Eurasians, Oceanians and Amerindians may have up to two per cent of their DNA from the Neanderthals, thanks to what is euphemistically called an 'introgression event'[‡] in the Middle East around 60,000 years ago. Melanesian groups in Oceania may also have up to five per cent of their DNA from Denisovan

***

[*] A paper published while I was writing this chapter shows that brown bears also have a teeny bit of cave bear DNA in their genome today. Around two per cent, or equivalent to how much Neanderthal DNA modern non-African people have.

[†] Denisovans are a complete mystery. We have their genome, thanks to a sliver of finger bone and a chunky tooth, but zero idea of what they looked like. No skull, no long bones, no definite associated material culture. All we know is that some of them lived in Denisova Cave in southern Siberia, and they were on the same evolutionary branch as the Neanderthals.

[‡] Sex.

introgression events. Thanks to the spread of shared DNA in the genomes of modern humans and the archaic species, we can even make some inferences about how the hybridising occurred. Our love of prurient details has no time limit. For the Denisovans, it probably involved Denisovan males mixing with sapient females, as there are fewer genes shared on the X chromosome. For Neanderthals, the same seems to have applied. Alternatively, there could have been some incompatibility that caused sterility in hybrid males. Either way, it's pretty staggering that for some lucky people, they have nearly a tenth of their genome from uniquely different human species. It's like finding aliens in your family tree. And it's not just super-sensuality on the part of modern *Homo sapiens*. The genetics show that Neanderthals and Denisovans were also 'introgressing' with each other. The Late Pleistocene has been described as like something out of *The Lord of the Rings*, with different intelligent species sharing the same world. Based on this evidence, I would suggest it was more like *Star Trek*, where inter-species 'introgression' was the norm. At least it was when Kirk was around.

I think it's an important point that when we think about the extinct megafauna of the Pleistocene and their interactions with humans, we are potentially talking about multiple different species. The way Neanderthals saw mammoths would have been different from how *Homo sapiens* saw mammoths, and from how Denisovans saw mammoths. And these three species, especially given their overlapping genes, can all comfortably be called human. All the evidence we have, especially from Eurasia, of human–animal interaction is necessarily vague, apart from a few sites where dating, stratigraphy and taphonomy (the science of how archaeological sites are formed) have given us privileged insight. Nowhere is this more apparent than in the evidence of contact between brown bears and people. Two examples illustrate this well.

The site of Regourdou is located near Lascaux in France. The ground here is riddled with caverns, and Regourdou was a collapsed one, evidence of geological processes still ongoing. The owner of Regourdou originally thought that he had found an alternative entrance to Lascaux, potentially a profitable discovery in the days when visitors were given entrance to the site for a modest fee. Instead, he had stumbled upon something just as intriguing, if less profitable: the skeleton of a Neanderthal. The skeleton, more or less complete except for the skull, was of a young right-handed male.* Positioned foetally, with legs tucked under chin, the Neanderthal was carefully placed within a stone-lined kist with an enormous capstone completely covering the grave. Replacing the shin bones of the man were shin bones of a brown bear. Adjacent to this was another stone-lined box, with the skeleton of a brown bear inside, mirroring the positioning of the man. Is this some kind of glimpse into the cultural beliefs of the Neanderthals? Can we see some conception of an afterlife by close association with the brown bear – a species that hibernates

---

* Amazingly, we can tell this based on both the asymmetric development of the arm bones (right is bulkier, from greater use) and from the orientation of the scratches on his teeth (right oblique, rather than left oblique). When using a toothpick, or cutting food held in his mouth, his dominant right hand left marks with the implement, giving away which side it came from.

in the winter and awakens to new life in the spring? Well, no. Everything I've said about Regourdou's bear cultism is total nonsense. And it's total nonsense that was believed by many archaeologists, thanks to the shoddy excavation that was originally performed. The original excavators really did not do a good job of keeping a detached and rigorous approach to what they were digging. Understandably, perhaps, they got lost in their own myth-making and failed to consider plausible alternatives, poisoning their site with speculation. A more level-headed approach to Regourdou suggests that while the Neanderthal was probably deliberately buried (shown by the completeness and anatomical association of his bones), there is no implication of any kind of relationship to a bear cult. Although Neanderthal expert Dr Becky Wragg Sykes, whom I quizzed on Regourdou, rightly points out that he could simply have been a sick individual left to die whose body either froze or got covered quickly to prevent scavenging. The original Regourdou Cave was used as a bear den: all those bear bones came from hibernating bears who didn't make it through the winter. The stone grave and capstone and the placement of bear bones are all accidental results of cave-ins.

How archaeological sites form is a big deal in our understanding of the past. Time is a tricky thing. All is flux. Nothing stays still for long. What can seem deliberate and conscious can actually be the result of thoughtless processes acting over geological time periods. In Regourdou, the collapse of the cave roof and the infilling of the cave over centuries gives an explanation for the imaginary stone coffins. They are just pieces of cave roof that have fallen, dropped and rolled into a facsimile of a kist. The stone walls and slabs imagined by the excavators were essentially rubble onto which they projected their dreams. The association of the Neanderthal bones with bear bones came from their displacement through water action, ground movement, even the presence of rabbits tunnelling through the earth. Good archaeology is not just piecing together the exciting aspects of a site and weaving a story but gaining an understanding of what happened both before and after. In cases like Regourdou, what was going on in the cave before the young Neanderthal died, and what happened in the cave in the centuries after his death, all had direct bearing on our understanding of his life. Regourdou aside, there really is almost nothing that ties Neanderthals and brown bears together. A few cut-marked bones here and there but that's it.

It's a very different story with our own species, *Homo sapiens*. Although we evolved in Africa, a place famously devoid of bears today,[*] upon entering Europe around 40,000 years ago we clearly found some affinity for brown bears. There is a beautiful mammoth ivory carving of a spread-eagled bear from the same region as the *Löwenmensch*. Wonderfully natural-looking brown bears crop up in cave art too. People knew and respected brown bears that shared their new homeland. Yet, our relationship with bears is strange. There is probably not a person reading this who didn't have a teddy bear when they were younger. Or read about Winnie the Pooh and

[*] A very recent state of affairs. The Atlas bear, *Ursus arctos crowtheri*, roamed North Africa until at least the nineteenth century. It may even have been exported to Rome for the games.

Paddington. I always think of Baloo from *The Jungle Book*, palling around with Mowgli and behaving like a surrogate parent. It's an easy contrast between these more modern images and older stories that still retain a hint of menace: the bears that pursue Goldilocks in the original fairy tale, the bears that eat 42 children in the Bible,* the bears that lurk in deep, dark caves, in deep, dark woods, waiting to eat you.

This oxymoronic love/hate relationship with bears does go back a long way. In Mesolithic eastern France, around 6,000 years ago, modern humans used a rock shelter near the Vercors Massif and left their waste flint and bones behind. In amongst the rubbish so gleefully uncovered by modern archaeologists were remains of boar and roe deer, ibex and chamois. Incongruously, there was also the jaw of a brown bear. Now, this wasn't just any brown bear. The mandible uncovered at this site attests to something unknown anywhere else in the archaeological record before this time. Between the first and second molars on the left and right side is a neat, circular depression. You could be forgiven for thinking that this was caused by some disease, although it would be weird for it to be so symmetrical. It's not from any disease we know of, and the bone looks healthy (given it has been in the ground for 6,000 years). What these symmetrical, round depressions are a sign of is a device that would have looked like a horse's bit, except

---

* 2 Kings 2:23–24 The kids had called the Prophet Elisha whatever the Aramaic version of 'a baldy bastard' is. Elisha curses the kids and two bears appear and maul them. The subspecies would have been *Ursus arctos syriacus*, which survives today in Iran, Iraq and Turkey.

made of wood or leather. This brown bear was tamed! About 6,000 years ago a Mesolithic human took a young bear cub and gave it a bit before the age of six months. We can guess the age because the first molar in bears appears at four months old, and the second molar at seven months old. The bit must have been in place between these ages to allow the teeth and jaw to grow around it. The tamed bear lived until about six years old, judging by the number and state of the teeth in the jaw. How mad is that! Think of what taming a brown bear would involve. The hunters would have to find a mother with cubs. They would have to, almost certainly, kill the mother to get those cubs. No easy matter when you consider that the maternal instinct in bears is so strong that it has passed into legend. Having obtained the cubs, the hunters would then have to find enough food to not only feed themselves but a rapidly growing bear cub too. When malnutrition and starvation were a constant threat, this represents a considerable investment. An investment paid out over at least six years. Bears must have been very, *very* important to these people. In wanting to tame the bear and be close to it, they must also have known how to hunt and kill it. Relationships don't get much more complicated than that.

Brown bears are tenacious survivors, despite everything we've thrown at them. They are smart, dangerous and mobile. How they coped with the climate change of the Pleistocene is a salutary lesson in adaptability. By tracking the bear lineages recorded in the DNA of their fossils, we can discern an interesting pattern. Today in Europe there are three distinct surviving branches of the brown bear tree: one found from the Spanish mountains to the

Swedish fjords, one from Greece to northern Finland and one in Italy. What's intriguing about these bear lineages is that there is very little overlap. Sure, they rub up against each other in certain places, but it seems like they are a signal of expansion.

You have to imagine that at the height of the Ice Age, most of central Europe and all of northern Europe was uninhabitable to most temperate mammals. Things either moved south or died as the glaciers moulded themselves from the mountains and gushed down the valleys to the coasts. While all of northern Britain was covered in ice, other sheets were forming in the frigid Pyrenees, in the Alps, and in the high Balkans and Carpathians. These walls of ice practically sealed everything in, preventing animals like brown bears from moving out of the Iberian Peninsula, the Italian Peninsula or the Balkan Peninsula while the ice was there. The bears were hemmed into these places for long enough that they evolved distinct lineages, readable in their DNA. When the ice began to melt, allowing expansion of the bear populations, they mushroomed outwards, only stopping when they bumped into bears from other refugia. This bottlenecking effect has left a signal in the genetics of brown bears and a few other European species.

This kind of study can even tell us something about the dynamics of British and Irish bears, using the genetic data that Ceiridwen, myself and other colleagues have generated. Those polar-loving Irish bears survived in Ireland up until the early Holocene, around 10,000 years ago. In this case Ireland may have acted as a refugium for that lineage of bears, even at a time when most of northern Britain and Ireland were covered in ice.

Meanwhile, a different lineage was found in northern England, before and after the Younger Dryas (the time that saw the demise of most of the shelk, see Chapter 6), suggesting a higher degree of adaptability in the bears than might be expected. The northern English lineage even went on to be found in mid-Holocene Ireland, about 5,000 years ago, when the polar-bothering lineage

Map 8: UK sites mentioned in this chapter.

had disappeared. They could have travelled from Yorkshire to Donegal without getting their paws wet, or at most having to do a little paddling. All the water sucked up into glaciers would have lowered sea levels so much that the Irish Sea was more like the Irish Puddle.

Brown bears and polar bears weren't the only Ice Age ursine animals. While brown bears do den in European caves, for a long time people told tales of something much more dangerous lurking in the dark. Here be dragons! Stories of gallant knights fighting mythical creatures are very common in European literature. The Lambton Worm, St George, *Game of Thrones*. Like all good myths, there is a hint of truth at their very base. It's not much of a truth, but it is instructive, and tells us a bit about how people interpreted fossils in the past. Terrifying European dragons are very different from the lucky dragons of China, and are most often said to haunt dark and dangerous caves, where they hoard treasure or steal people and livestock to eat. It's generally thought that these legends come from people finding caves full of real but fantastic bones – the bones of extinct creatures that were unlike anything people knew. Take the dreaded Lindworm, an Austrian dragon. This giant reptile flooded the town of Klagenfurt and threatened to eat the sodden inhabitants. The king put a price on its head and it was slain by a group of enterprising local knights. Klagenfurt has a statue of the slippery Lindworm, erected in the sixteenth century, in honour of the historic victory over its magic. The head of the statue is based on the skull of a woolly rhinoceros, found in the local, fossil-rich caves. You see, in the dark caves, seldom explored, people conjured up dragons and trolls, Lindworms and monsters.

Dragons are some of the first attempts at explaining fossils, and the Lindworm of Klagenfurt is the first example of palaeoart. Europe's caves are jammed full of Pleistocene bones.* This is as true in Britain as it is in Austria. The vast, vast majority of those bones come from one species in particular: the cave bear (*Ursus spelaeus*).

Cave bears are awesome. Their closest living relative is the brown bear but they were substantially different. Cave bears were absolutely enormous, for a start. Estimates have large males weighing up to 1,000 kilograms (2,200 pounds). That's twice as heavy as the average grizzly. Cave bears packed on all that bulk with a completely vegetarian diet. Think of them as like a native European analogue to a giant panda† – a (mostly) gentle giant. We know from the stable isotopes that cave bears only ate plant material. The evidence is also written in their teeth and skulls. The cave bears have lost some of their sharp premolars and transformed their true molars into flat, grinding surfaces that intersect like pestle and mortar to pulverise plant remains. The skull is distinctively different from all other bears as well as being much bigger. The air-filled sinuses are massive and give the

---

* So many bones! Some caves were actually mined during the First and Second World War for their bones. The hard part of bone is a mineral called apatite that is mostly calcium and phosphorus. The phosphorus content of the bones was put to use in munitions.

† Giant pandas are really also part of the bear family, and the most basal member. Like the hyaenas, the bears are a small lineage. Alongside the giant panda there is the spectacled bear, sloth bear, sun bear, black bear, moon bear, brown bear and polar bear.

animal a primitive, bulbous forehead that is faintly comical. However panda-like in diet, the bones suggest that they may have been less cuddly in life. Amongst the millions preserved, many bones show evidence of serious damage. Healed bite wounds, fractures and breaks show that life was tough for the cave bear. Either they were doing a lot of fighting amongst themselves or other species were actively hunting them. How do you hunt a literal tonne of bear?

Well, it probably ties in to why our caves are so full of their bones. Cave bears were hibernators. Every autumn, cave bears, young and old, would put down a layer of fat as best they could to survive the bitter cold of winter. Moseying back to their caves, they would shut down for the season in the dank recesses. For non-hibernating hunters, like the cave lion and cave hyaena, these winter caves would represent an all-you-can-eat buffet. Puncture marks corresponding to the teeth of these carnivores are common in cave bear bones.* Cave bears that didn't fall victim to a predator's jaws wouldn't necessarily see the spring. The very young and the very old are over-represented amongst the bones – signs that for the feeble and the weak, not eating enough in the autumn was a delayed death sentence. Their slowed metabolism during winter would still burn through all

---

* Many sources claim that the oldest musical instruments are a series of flutes found in Slovenia and Hungary and made by Neanderthals from the leg bones of young cave bears. The bones have multiple perfectly round 'finger holes' and look like they were made for playing. Appearances can be deceptive. Those holes come from cave hyaena canines punching through the soft baby bone.

their fatty fuel, starving the animals to death in their sleep. This brings up an interesting conundrum that is common to all of palaeontology: we only see the ones that didn't survive. Our understanding of cave bears and many other extinct species relies on making inferences about the dead and not the living. Usually, the assumptions hold, but it's always worth keeping in mind that nearly all our fossil bones come from individuals that didn't quite make the grade.

Nevertheless, unlike with some species, we have an abundance of riches with the cave bear. Finnish palaeontologist Björn Kurtén[*] made the study of cave bears via statistical methods his life's work and wrote the definitive book on them – *The Cave Bear Story*. Thanks to him, and later scholars, the cave bear has really been at the vanguard of research on Pleistocene megafauna. For example, cave bears were the very first extinct species to have their nuclear genome sequenced, way back in 2005. They were a natural choice given how many well-preserved bones were available to researchers.

Close study of the morphology of the bones also allows some interesting conclusions to be reached. First of all, not all cave bears are *Ursus spelaeus*. Bimodal size differences give an impression of sexual dimorphism – that is, males were much bigger than females. But shape differences give us a picture of different kinds of bears in different times and places. For historical reasons, these differences

---

[*] He wrote in his 1976 book *The Cave Bear Story*: 'No one escapes his fate. It might be said that my affair with the cave bear started half a century ago when it was decided to give the child a name that happens to be Swedish for bear. There were some early difficulties in living up to it.'

have been used to place different populations into different species. Whether they actually were different species is a particularly thorny problem. Traditionally, biologists define species as animals that normally breed with each other to produce fertile offspring. Horses and donkeys are different species because their offspring (mules) are fabulously sterile.* There are other, competing ways of defining species, including by DNA analysis, and these are as arbitrary as any attempt to impose rules on nature. With the different forms of cave bear, we run into an immediate problem. Nobody's gonna be breeding with anybody cos everybody's extinct. Some of those cave bear species were given names: *Ursus deningeri*, *Ursus rossicus*, *Ursus ingressus*, *Ursus eremus* and *Ursus ladinicus*. Within this species complex, some broader patterns seemed to emerge. *Ursus deningeri* was the oldest form,† found in the most ancient caves dating back to the Middle Pleistocene. Britain has a particularly rich record of *Ursus deningeri* in the caves of Kents Cavern, Westbury-sub-Mendip and Wookey Hole. It was the direct ancestor of some of the later cave bear species including *Ursus spelaeus*. *Ursus deningeri* may have survived in an isolated pocket of the Carpathian Mountains, separated from the other cave bear populations it gave rise to. On the other hand, *Ursus rossicus*, *Ursus eremus* and *Ursus ladinicus* were likely just derived

---

* Not always. Everything in biology has an exception and there are a number of records of fertile mules going as far back as Herodotus.

† For a while the record for oldest DNA ever recovered was from 300,000-year-old Middle Pleistocene *Ursus deningeri* bones from Spain. This record has been broken several times.

populations or subspecies adapted to local conditions. These 'species' are typically found in caves at much higher altitudes than *Ursus spelaeus* and tend to be smaller, perhaps as an adaptation to the high alpine climate. The only other form defined by the bones that seems to be legitimate is *Ursus ingressus*. Whether it was a highly derived population, subspecies or even species is impossible to say from just the bones, but it looks like it first appears in eastern Europe and then spreads into western Europe during the Late Pleistocene.

Thanks to ancient genetics, it's become possible to test some of the hypotheses based on the study of cave bear bones. It's simple in theory: a question of trying to correlate particular forms found in the bones with particular sequences found in the DNA. If the two match up, then that shows us that the distinctions are valid. They pretty much do but with some interesting details. The dwarf forms of cave bear (*rossicus*, *eremus* and *ladinicus*) actually have very small distributions and are only found in a few high-altitude caves.

There is also a noticeable lack of continuity between different caves really quite close to each other. Ramesch Cave and Gamssulzen Cave in Austria have remains of small *Ursus eremus* and large *Ursus ingressus*, respectively. For over 10,000 years the two caves retained entirely separate DNA lineages, despite only being a few miles apart. This could mean that some of the morphological forms do represent distinct species and were purposefully not breeding with each other due to differences in mating cues. This is especially weird since the lineage in Gamssulzen, the large *Ursus ingressus* type, is found all over Europe. Perhaps some of the high alpine habitats were evolutionary islands, reproductively isolating cave bear populations for long periods of time and allowing them to evolve unique forms that kept them separate from more cosmopolitan relatives. Similar things happen in other parts of the world, particularly rainforests punctured by high-altitude mountains. These 'sky islands' can be as isolated from each other by dense foliage as oceanic islands are by water. The inability to move easily from mountain to mountain is enough for evolution to work its magic and start the process of speciation.

The large *Ursus ingressus* form of cave bear also has another facet apparent in the genetics. Comparing radiocarbon dates and DNA tells a story of stark population replacement. Around about 28,000 years ago the lineage of the classic *Ursus spelaeus* type abruptly disappears in western Europe. *Ursus ingressus* magically appears with almost no overlap. It's like it has been rubbed out and drawn back in. It's hard not to see the hand of humanity in all of this. Cave bears had coexisted fairly peacefully with Neanderthals for the better part of a million years – using the same caves, if not at the same time, and spearing the occasional bear. It's a different tale after *Homo sapiens* came waltzing into Europe. Then, we see a very different pattern. Our species seems to have excelled at killing cave bears, by necessity and by design. There is an abundance of cave bear bones with cut marks, and some still retain the flint points that killed them. An example from Hohle Fels Cave[*] in Germany is particularly instructive. This cave bear vertebra has a triangular flint point stuck in one of the lateral processes,

---

[*] The same cave that gave us an impressive Venus figurine.

the spurs of bone that back muscles attach to. The angle of entry shows that the cave bear must have been lying on its left side with the hunter jabbing down into it from above. This wound couldn't have come from a thrown weapon – the angle is wrong; it must have been a close-contact kill. The only way to get that close to a cave bear would have been during hibernation. Picture the scene: a nervous and skittish hunter braving the black cave depths to find their quarry, senses attuned to the dark, listening out for the soft snuffling of a hibernating monster. Would the use of a small oil lamp or torch for light cause the monster to wake? Agonisingly slowly, the hunter creeps up to the tonne of sleeping bear and hesitates before plunging a wood and flint weapon deep, deep into the flesh. On this occasion the hunt was successful, attested to by cut marks on the vertebra. In the harsh world of the Late Pleistocene, this might have been how modern human groups made it through winter after winter.

Amazingly, there are signs that this kind of hunting was drilled into young learners with all the precision of a military boot camp. Montespan is a cave site in the Pyrenees. Difficult to access thanks to underground rivers, it was entirely cut off from the outside world until an enterprising and thrill-seeking young speleologist swam through its black waters. Surfacing on the other side, he found that he was not the first person to visit this particular cave. Evidence of prehistoric people was daubed all over the walls. More impressively, the Pleistocene artist who had used this place as a studio had sculpted the body of a bear out of clay on the floor. Riddled with holes from spearpoints jabbed into the side, a bear skull sat between its front paws, sphinx-like, maybe pointing towards the idea that the clay body had originally been covered in a head and skin. Perhaps the sculpture had been used to show inexperienced hunters how to strike hard and true to kill a sleeping bear. What kind of bear remains a frustratingly unanswerable question. Although brown bear and cave bear skulls are very easy to tell apart, this unique relic was stolen from the cave before any experts had the chance to lay eyes on it. Hibernating cave bears could have been an abundant source of relatively easy food during winter. At such stressful times every part of the animal would have been used, not just the meat. Obviously, the fur and pelt provide warmth and can be used for clothes and leather. Teeth were cut from the jaw for use as jewellery and ornaments. The bones themselves were burnt for fuel. Like with Native American groups living with bison in the Americas, every part of the cave bear seems to have had a use. Even the baculum, the penis bone* found in many mammal species, shows evidence of specialist use in the cave bear remains from Hohle Fels. The wand-shaped bone has the sheen of polish that comes from hours of working hard skin into good leather.

The cave bear's importance as a food source would have easily incorporated it into the complex cultures of Ice Age humans. It used to be thought that a cult that venerated and worshipped the cave bear existed even

---

* Most mammalian species have a penis bone, the baculum, and a clitoral bone, the baubellum. Humans are oddities in not having one, since our closest relatives the chimps and gorillas do. Walruses have one of the biggest, with an enormous bacula 0.5 metre (1.5 foot) long, used by native Alaskans culturally and called 'oosik'.

amongst Neanderthals, but this idea has been discarded due to dodgy excavations and weak associations.[*] Still, there are some incontrovertible signs that cave bears held an important place in the pantheon of modern humans who replaced the Neanderthals. Cave bear bones are occasionally found covered in red ochre,[†] a symbolic pigment that was also used to colour buried bodies in the Pleistocene. Immediately, this suggests a link between the treatment of dead humans and dead cave bears. Artists painting on cave walls also used red ochre. It's difficult to really differentiate between cave bears and brown bears in cave art; although there were differences in size and appearance, these don't translate well to two-dimensional images without any scale, especially since we don't know what *Ursus spelaeus* looked like in life. However, there are a couple of dead certs in Chauvet Cave, host to many of the best images of that era. They show the remarkable bulbous forehead we can see in the skulls. The images are very vivid, outlined in red ochre, like the bones. They do a good job of showing the shared humanity between human and bear; the curious snout thrust forward, round ears attentive. Chauvet was also a cave bear den and has an

[*] This idea led directly to one of the most vivid fictional depictions of Ice Age life ever published, in Jean M. Auel's *The Clan of the Cave Bear* book series. If you haven't read them, you must. They are full of incident informed by real archaeology, and lots of hot, hot human-on-Neanderthal action.

[†] Found in a range of Belgian caves, including Goyet, famous for its Neanderthal bones.

abundance of remains.* A cave bear skull was found placed on a rock pedestal by human hands, looking for all the world like an altar and Eucharist. Charcoal stripes were deliberately placed on a different skull, for reasons unknown. Bones of cave bears have been placed in fissures in the wall by human hands. It reminds me of the devout slipping prayer notes into the cracks of the Western Wall. Who knows why people put bones where they did? Prayer or boredom, devotion or play, it will be a mystery for years to come.

The remains of hibernation nests are still to be seen in Chauvet Cave. A pathetic reminder that cave bears are long gone, the empty nests would have been scratched out of the soil before hibernation, the bears twisting and turning to get comfy before settling down for the long winter. The walls of Chauvet are full of scratches from the claws of cave bears too. They sharpened their blunt claws as modern bears do. Sometimes the scratches are over the top of paintings. Cave bears as art critics. Mostly the claw scratches are on bare patches of cave or underneath the masterpieces daubed on the walls. Perhaps all of Chauvet's art started as an attempt to mark the cave as a place for humans, just as surely as the cave bear scratches had marked it as a place for bears. Footprints can be found too, those from cave bears dwarfing those of people, all pressed down into the Palaeolithic mud and hardened to stone after 300 centuries.

---

* Really well preserved too. The first complete cave bear mitochondrial genome was sequenced from a bone from Chauvet in 2008.

Cave bears are gone and brown bears are here, which poses a slight conundrum. These are sister species, albeit with some clear differences ... so why did one disappear and the other survive? I think there are enough clues to give us some idea of what actually went down. We know that cave bears were less flexible than brown bears in some of their behaviours. Genetic comparisons between the two species in caves in northern Spain throw up an interesting difference. For thousands of years cave bears of the same lineage came back again and again to the same caves, whereas brown bears of different lineages are found in multiple caves. The conclusion is clear – cave bears were tied to their hibernation sites and returned year after year, their cubs inheriting the same cave and unthinkingly following their ancestors. On the contrary, brown bears had the flexibility to move cave on a whim, or if presented with an obstacle. In a similar vein, cave bears seem to have been strictly vegetarian, according to their stable isotopes, or at best only occasionally omnivorous. Brown bears had the behavioural flexibility to switch between carnivory and herbivory based on what was available.

This ability to shift with the situation is a great adaptation for survival in an uncertain world. Evidence shows that modern humans hunted hibernating cave bears. If the poor bears had no choice but to come back to the same caves year upon year, and generation after generation, then the culling could not fail to have had a serious effect. In the uncertain and variable climate of the Late Pleistocene, having strict dietary requirements would be an additional burden to survivors. It wouldn't take much pressure to push a species like the cave bear to extinction. Think of how the giant panda has fared with humans over the past few centuries. Very few people have tried eating them,* but the clearance of bamboo forests has forced them into ever-more marginal habitat where they struggle to produce enough offspring to replace themselves. Cave bears could have been similar: harried by people butchering them in the winter, struggling to find enough food in the summer, in conflict with people all year round for use of caves as shelter. Getting whittled away. This is exactly the pattern we find recorded in their population structure using a comparison of radiocarbon-dated and DNA-sequenced individuals to reconstruct population change (similar to what was done for bison, see Chapter 7). Starting around 45,000 years ago, (precisely the time that modern humans arrive in Europe from the east), the population size of the cave bear goes into a steep nosedive. At the same time, brown bears remained at a steady population size. The decline is so stark that even back in the 1960s, Björn Kurtén, using just the bones, was able to identify 'that the extinction of the cave bear was not a sudden thing but a gradual process, spread out over several thousand years'. The very last cave bear we have a date from anywhere on earth is 19,686 radiocarbon years old (about 24,000 calendar years old). It's from a cave in northern Italy.

I may be an old romantic, but for me the saddest, most poignant remnant of the time of the cave bears is so

---

* President Roosevelt's sons Kermit and Theodore Jr are on record as the first westerners to shoot a giant panda, for the Field Museum in Chicago. Of course they ate some of it (their guides were horrified) but did not record how it tasted.

subtle that it escaped notice until fairly recently. Within many caves, all over Europe, there are weirdly polished surfaces in nooks, crannies and tunnels that defied explanation. These surfaces are *'Bärenschliffe'* – literally, 'bear-shine' – caused by the rubbing of tens of thousands of bears as they entered and exited their hibernation den year after year. It's like the answer to some Zen koan of how to define infinity: if, once a year, the fur of a cave bear touches the rock of the mountain wall, when the mountain has eroded completely, one second of infinity will have passed. It is an infinity that will never come.

CHAPTER NINE

Northern Lynx

Extinct in Britain after the
seventh century AD (medieval)

> Lions we haue had verie manie in the north parts of
> Scotland, and those with maines of no lesse force than they
> of Mauritania were sometimes reported to be; but how
> and when they were destroied as yet I doo not read.
>
> Raphael Holinshed, *Holinshed's Chronicles of England,
> Scotland, and Ireland* (sixteenth century). Possibly
> describing the extinction of the lynx from Britain.

I have a small confession. Despite spending 20 years of
my life looking at the genetics of lions and other big
cats, I've never seen one in the wild. I've never been to
Africa or India, where lions cling on today. The closest
I've come to Africa is staring at the Atlas Mountains in
distant Morocco from the Rock of Gibraltar, imagining
how much it has changed since Barbary lions roamed
free.

I've had precisely one close encounter with a wild
felid, and the few seconds it lasted will be stamped on
my soul until the day I die. It was back in 2003, when I
was roving from gold mine to gold mine with colleagues
looking for fossil bones in Yukon. The distances
between mines can be huge, and we had hired a rugged
4x4 to ferry us around. In the back of beyond, a couple

of hours' drive from Dawson City, I had my encounter. As we rounded a bend, about 30 metres (100 feet) ahead of us, a lynx was waiting on the road. Anyone who has spent time with any kind of cat knows that apathy is their leitmotif. This lynx was the king of indifference. It looked nonchalantly towards us without any visible signs of recognition. Then it continued, shoulders lolling, to the other side of the road. What I remember is how big it was. People think of lynxes as 'medium-sized', but this guy looked at least the size of a Labrador. Big enough to make you think thrice about trying to pet it. Long-legged, fluffy to the tips of its ears and cheeks and with the short tail only found in lynxes, this cat was effortlessly at home in Yukon. We rushed over to where it had left and saw it unhurriedly ambling away amongst the spindly trees. Off to do secret lynx things that didn't concern the likes of us. It was only a few moments but it will remain with me forever.

Lynxes are special to me. In the tangled web of felid phylogeny, they stand as a coherent group. A gang of four: the bobcat,* the Canada lynx, the Iberian lynx and the Northern lynx. They are a resolutely northern-hemisphere bunch. Fossils suggest that they may have originally evolved in North America, where the bobcat and Canada lynx live today. Africa used to have lynxes, shown by fossils, although none live there now. Instead,

---

* I always wondered why bobcats were called bobcats. It's because they have a short, bobbed tail, like all lynxes.

Africa has the caracal,* sometimes known by the taxonomically incorrect name of 'desert lynx'. Skeletally, lynxes and caracals are very different: caracals have a long tail, for a start. They do have very lynx-like ear tufts, though.

All four of the modern lynx species can trace their evolution back to one long-lived and widespread cat known as the Issoire lynx.† The Issoire lynx lived between the Late Pliocene and Middle Pleistocene, around three to 0.5 million years ago, and from Spain to China. First to split off from this branch was the feisty bobcat. Smallest of the lynxes, it's still at least twice as large as a big Maine Coon domestic cat. Bobcats are found all over the US and Mexico, even creeping up into Canada. They are notoriously fierce, despite being small, and are not to be trifled with. Nevertheless, archaeological sites in Illinois show that they were at least occasionally kept as pets by Native Americans, and buried with due ceremony.

The next offshoot of the lynx family is the Canada lynx, the species I had the privilege of viewing in Yukon. Their coat is the thickest, to cope with the cold, and

---

* Caracals are also found in the Near East and India. In the past, tame ones were kept as pets. The origin of the phrase 'to put the cat among the pigeons' comes from the Indian practice of de-bagging these tamed cats into a flock of grounded pigeons and betting on how many the dexterous animals could take down. Particularly nimble individuals could claw down dozens before the flock had made its way into the sky.

† Because this is the region of France where the fossils were first found. Pronounced 'ees-whar'. Isn't that a fabulously sensual name? It just slinks off the tongue.

they even have hair on the undersides of their paws to insulate them in the snow. Their feet are enormous too, acting as natural snowshoes.

The final two species to split off from the lynx family are the Iberian lynx and the Northern lynx. They've had a hard time recently. Europe's only endemic cat species, the Iberian lynx, came within a whisker of being completely wiped out. To put that into proper context, no species of cat has gone extinct since the end of the Pleistocene. The Iberian lynx is perilously close to joining the cave lion and the scimitar-tooth. Just a few hundred survive today. In 2002, at what I hope was the climax of lynx destruction, there were only 52 mature adult cats left in the wild. Conservation work has been intense, and has helped to breed and release captive individuals at a number of sites. At the beginning of the twentieth century, the Iberian lynx was found all over Spain and Portugal, from the Pyrenees to the Algarve. Now its last stronghold is the Doñana National Park in Andalusia, with a few more holding out in the Sierra Morena.* The quick decline in population numbers is a direct result of hunting, habitat change and the devastating effects of myxomatosis (a fatal viral disease) on rabbits, the prey item that makes up between 80 and 99 per cent of the Iberian lynx's diet.

It used to be thought that the Iberian lynx was just a subspecies of the Northern lynx; however, there are clear genetic and morphological differences. The Iberian lynx

---

* In November 2007 a third, previously unknown population of around 15 animals was found in Castilla–La Mancha. Quixotically, the regional government has decided to keep the exact location of their lynxes secret to protect them from hunters.

is much smaller, more spotted and with more facial hair. The clincher for separating them comes from a careful study of the genetics of their respective fossils. During the Pleistocene, the Iberian lynx had a much wider range and could be found even into northern Italy. At the same time, the Northern lynx had spread south into Spain, allowing the two species to overlap in parts of France, Italy and Spain. Despite the overlap, there is no sign that the two species ever hybridised, suggesting very strongly that they are real and distinct biological species. This overlapping went on for a long time, even into the sixteenth century AD when the Northern lynx could still be found in Cantabria.

The Northern lynx has not been doing well in Europe either, and although its decline has mirrored that of the Iberian lynx, it's been happening for a lot longer. At their peak, Northern lynxes were found in northern Spain, Italy, France, Germany, the Low Countries and even Britain. It used to be thought that lynxes had disappeared from Britain at the end of the Ice Age, tying their extinction neatly into the same 'mysterious' forces that killed the woolly mammoths and rhinos. In fact, that's not the case at all. Lynxes lived in the British Isles for 10,000 years after the rest of the megafauna disappeared. The last radiocarbon date for a British lynx comes from an animal that was alive in the late fifth or early sixth century AD. This animal was found at the site of Kinsey Cave in Yorkshire. Another lynx, dated to a couple of centuries prior, came from Reindeer Cave at Inchnadamph* in Sutherland. Given that no radiocarbon

---

* You'll recall from the bear chapter that this site is rich in other extinct species as well.

date is likely to ever be obtained on the last member of
a population, they were certainly around later than this.
To put it into context, lynxes were definitely in Britain
later than the Romans and were still around when the
Saxons arrived. So, since their survival lasted solidly into
the historical period, what was it that did them in? It
clearly wasn't climatic factors, as nothing particularly
horrendous is recorded in chronicles written at the time.
Lynxes in the rest of Europe survived just fine too. It
must have been local, human pressures. Around this time
there was a lot of political turmoil, with many small
kingdoms warring with each other, and lots of intense
deforestation making way for agriculture. It seems
certain to me that a combination of direct persecution
along with destruction of habitat caused the extinction
of the lynx in the British Isles.

Now, lynxes have one notable advantage over most of
the other species I've discussed. They aren't extinct, yet.
There are large, healthy lynx populations in Scandinavia,
Russia and in parts of Eastern Europe. Their second
advantage is that they don't eat people. Unlike with
bears and wolves, nobody has ever been killed by a lynx.
Worst-case scenario if you rile up a lynx, you're going to
need plenty of Steri-Strips, antiseptic and a new attitude
towards messing with wildlife. Lynxes are an ideal
candidate for reintroduction into Britain. As a species,
they've sidestepped the killer reputation that is drummed
into us from the age we first listen to fairy tales. They
also look adorable and fuzzy. Lynxes are the best bet for
rewilding with a lost native carnivore.

In a way, the experiment has already happened. And
the results were not pretty. Back in late October 2017, no

Map 9: UK and Irish sites mentioned in this chapter.

one knows exactly when, a lynx held at Borth Wild
Animal Kingdom near Aberystwyth escaped. Lilleth, as
she was known, was only 18 months old and quickly
made the Welsh countryside her home. In the nearly two
weeks she was loose, pandemonium reigned. Dubbed
the 'Beast of Borth', Lilleth, an entirely hand-reared
animal, not fully grown and with no experience of living

in the wild, did her best to survive. She may have killed seven sheep. The animals were found, mostly uneaten, with classic felid neck bites, less than half a mile from the zoo where Lilleth lived. This carnage triggered a frenzied response from the NSA. No, not the National Security Agency, but the almost-as-terrifying National Sheep Association. Phil Stocker, of said organisation, went on record to say: 'There cannot be a clearer warning of the damage lynxes will do if they are released into the wild.' Lilleth managed to evade capture for a few days more. Tragically, she tried to hunker down in a caravan park in Aberystwyth and the council stepped in. Local marksmen shot her dead on the morning of Friday 10 November 2017, bringing her freedom to an end. Lilleth's case is a tragedy, initiated by lax enforcement of animal welfare rules in private zoos. Borth Wild Animal Kingdom was later banned from keeping lynxes or other dangerous creatures on their facility. It's a sad end, considering Lilleth never hurt any people and there was even some argument over whether the sheep had been killed by her or feral dogs. For the first time in many centuries, a lynx had roamed free in Wales.

This kind of thing happens reasonably often. People claim to see the 'Beast of Bodmin', the 'Surrey Puma' or the like, and occasionally there have been real cats behind the sightings. One of the most famous – and well-recorded – cases is that of 'Felicity the Puma'. Between 1978 and 1980, farmers near the wee village of Cannich, west of Loch Ness, reported unusual sheep losses that they claimed were from a lioness. One farmer, a local man called Ted Noble, went so far as to construct a specially built cage-trap that could be baited with raw meat, reminiscent of those used in publicised attempts to capture the Loch Ness Monster. Nobody was more shocked than Ted to actually find a female puma in the sprung trap on 29 October 1980. The puma, Felicity, was in good condition, friendly and used to people; reports have stated that she liked being tickled behind the ears. Analysis of her scat, produced directly after getting trapped, showed that she had been eating deer and rabbit – wild game. Felicity was resettled at the Highland Wildlife Park in the Cairngorms, where she lived for five uneventful years as a big tourist draw. When she passed away from old age, she was taxidermied and put on display at the entrance of Inverness Museum. I remember vividly seeing her there myself as a youngster. Pumas are native to the Americas, not Scotland. Felicity was a long way from home and there was no zoo within 160 kilometres (100 miles) that she could have escaped from.

I mentioned in the chapter on lions that the market for exotic animals is huge. Never was this more so than in the Swinging Sixties.* If you had the cash, you could walk into Harrods and buy a lion cub, no questions asked.† Unfortunately, as many people found out, large exotic cats are not easy to care for. Out of concern for public safety, the UK government brought in the Dangerous Wild

* By way of example, in the 1960s the surrealist Salvador Dalí had a pet ocelot named Babou. He also kept multiple anteaters including a giant anteater and a northern tamandua.
† This is exactly what happened with the famous lion called Christian: bought by a London couple, the lion outgrew the capital and was released back into the wild under the tutelage of George Adamson. Christian's story became inspiration for the film *Born Free*, and was one of the early viral hits for YouTube.

Animals Act 1976 to try to curb some of the previous decade's excesses. Anyone wanting to own an exotic pet would have to first apply for a licence, which would only be granted if the government were happy that the facilities to house it were adequate. Likewise, the act required liability insurance to be purchased to cover any potential danger posed to the public. For some who inhaled too deeply the anti-authoritarian views of the period, this legislation was too much. Under the guise of sticking it to the man, exotic pets were secretly dumped to prevent them causing financial or legal problems to their owners. Felicity's appearance in sleepy Cannich so soon after the passing of the act is no coincidence.

However, it's long been realised that sightings of weird cats in the wild parts of Britain didn't start in 1976. Lilleth isn't even the first lynx to have roamed wild in Britain since medieval times. I was part of an interdisciplinary study looking at one of the other verified cases – a cat that has since become known as 'the Bristol lynx': an Edwardian big cat, loose in genteel Devon. The project started when Max Blake, a student working in Bristol Museum, came across an old accession record for a taxidermied lynx, stating succinctly that it had been shot in Newton Abbot in 1903, after killing two dogs. Now, lynxes are distinctly not part of the regular Devon fauna, but somehow, this fact had not caused any discernible eyebrow-raising when the specimen was first curated.

It did for us. Max contacted Darren Naish, a palaeontologist with an interest in anomalous animals.*

---

* Darren writes extensively on palaeontology and cryptozoology on his blog *Tetrapod Zoology*. Get it read.

Darren got in touch with me and a host of other experts to throw a battery of tests at this weird lynx. I was involved with the ancient DNA analysis, others were looking at the morphology of the bones and the stable isotopes. Together, we hoped to identify what lynx species we were looking at, which region it may have come from and how long it had been in the wilds of south-west England. Annoyingly, despite giving it my best shot, I couldn't get any DNA from either the taxidermied hairs or the associated bone. The lynx was only a wee bit over a century old. I'm used to working on DNA from cave lions that are over 50,000 years old. The inability to get the damn thing to work was irritating.

On reflection, now that I've calmed down, the best explanation I can conjure up is that sometime during the preservation process of the pelt and bone, the curators treated them with something that interferes with all the specialist enzymes that need to work to amplify DNA. It's not so unlikely: Victorian curators were very liberal with the application of toxic arsenic powders and solutions as a way to prevent insect pests munching through museum collections.* Perhaps the Bristol lynx had received some application that stopped me from getting any DNA.

---

* Arsenic is no longer used in museums, thanks to elf and safety. The issue of insects eating into irreplaceable specimens still remains, though. Some museums freeze their samples to try and stop the plague of various moths and beetles that eat skin, hair and feathers. A continued nuisance in museums is the carpet beetle, whose fuzzy larvae have the cute nickname of 'woolly bears'.

Luckily, some of our other experts were able to get answers, and the proportions of the bones showed that it was a Canada lynx. The mystery deepened. Canada lynxes are even less native to modern Britain than Northern lynxes. The Bristol animal showed signs that it had been kept in captivity for a considerable length of time. All the incisors, the front teeth, had been lost and the bone had grown back in – a pathology that would have been fatal in the wild. The remaining teeth, particularly the molars, were totally covered in plaque. This lynx must have had someone who kept it fed and looked after when it couldn't have done so itself. A wild, natural diet keeps plaque in check. On the skeleton too, some of the toes had been rudely cut, something that happens during declawing. In cats of all species, it's not possible to cut the claw without the underlying bone being removed as well; declawing is essentially amputation, and that's what seems to have happened here. Overall, the story is of a captive Canada lynx that was released or escaped at the beginning of the twentieth century, because it was too old, too sick, too expensive to keep or just not needed – all valid concerns for owners of exotic pets, even well before 1976. It could have come from a travelling menagerie. It could have been the pet of a sailor, newly back from the Canadian territories. Whatever the truth, the Bristol lynx was unprepared for life in the wild. Its time living free in Devon was probably short. If it had not been shot by the farmer in retribution for killing his dogs, it would surely have died soon from starvation and been lost to memory. As it stands, the examples of Lilleth and the Bristol lynx show that lynxes can live in modern

Britain – it's just that it only takes one person to want them dead.

A lot of work has gone into examining the feasibility of bringing lynxes back. In Britain, they have been extinct for a millennium. In parts of Europe, amidst increasing persecution, many populations only disappeared in the early twentieth century. France, Germany, Switzerland, the Czech Republic and Slovenia have all stepped up and instigated reintroduction programmes for lynxes to replace native populations driven to extinction. One of the success stories involves the Dinaric Alps population. This mountain range stretches through Croatia, Slovenia, Bosnia and Herzegovina in the Balkans. It's thought that a hunter shot the last lynx in the region in 1911. In 1973, after 62 years of absence, an ambitious attempt to bring the lynx back was put into action. Six lynxes from the Slovakian Carpathians* had been captured and taken to Kočevje Forest in Slovenia. When the Dinaric population was exterminated, the Carpathians remained the geographically closest population source. In Slovenia, those six lynxes immediately started to reproduce, producing litter after litter, and spreading into Croatia, Bosnia and Herzegovina. Over 40 years later, the population now stands somewhere around 140 – all from six founders.

The Dinaric Alps reintroduction has been a phenomenal success by most parameters. However, there are some issues. Because the number of founders was so low, genetic diversity in their descendants is even lower, which has real-world ramifications. With low diversity

---

* Including a mother and son and a brother and sister pair. Not ideal in terms of genetic diversity.

populations, inbreeding becomes a big issue, and the greater the chance of inheriting two copies of a duff gene that could cause abnormalities. This problem has been looked at in another successful lynx reintroduction population from the 1970s, in the Swiss Alps and Jura. Unlike the Dinaric project, this one was done haphazardly and with minimal records. We don't know how many lynxes were translocated, though it was likely fewer than 30 in total throughout numerous Swiss cantons. Starting in the 1990s, reports of malformed and abnormal lynxes began to appear – lynxes with scoliosis of the spine, fatal hernias, monorchidism,* severe underbites and heart murmurs. Many of these issues had already been described as the results of inbreeding in a different species of cat: the Florida panther.

The Florida panther is the last subspecies of puma to survive east of the Mississippi. Due to human pressures, the subspecies had been reduced to fewer than 50 individuals before people began to notice that things were far from OK. Numerous Florida panthers were found with cryptorchidism,† kinked tails due to bone abnormalities, and other issues. They had entered what is known as an extinction vortex. Individuals had abnormalities due to inbreeding; they could only mate with other abnormal inbred individuals, leading to offspring with even more abnormalities, making them

* This lynx has only got one ball. Mono = one, and orchid = testicle. The orchid flower is named because the roots resemble a testicle.
† Crouching puma, hidden testicle. Cryptorchidism happens when one or both testicles fail to descend, leading to infertility.

even less likely to survive to reproductive age. It's called a vortex because once a population enters it, it gets sucked ever downwards until there are no survivors. This was to be the fate of the Florida panther, until some clever wildlife biologists stepped in and said 'not today'! To alleviate the problems with inbreeding, a number of pumas were taken from Texas and released in Florida. The injection of new blood into the gene pool had an immediate rejuvenating effect. In just a generation, the offspring of couplings between Florida and Texas pumas were seen to be more robust and more resilient than 'pure-bred' Florida panthers. Overall, the total population tripled. The story of the Florida panthers and their genetic rescue with Texas pumas is a conservation success story, showing what can be done to reverse the fortunes of isolated and inbred populations. Even for the inbred populations of Switzerland and the Dinaric Alps, recovery of diversity is possible by adding in fresh genes. In lynxes, though, genetic diversity might not be as essential as we used to think.

I spoke to Dr Ricardo Rodríguez-Varela, an expert on the genetics of the Iberian lynx, and the first person to identify Northern lynxes in Spain and Iberian lynxes in Italy by their DNA. He told me about one of his 'eureka' moments. Working on fossils of Iberian lynxes, Ricardo could reach back as far as 50,000 years, and what he discovered was super-interesting. It's been known for a long time that modern Iberian lynxes are very uniform, genetically. Hardly unexpected given their current tiny population size. Every living Iberian lynx has the same mitochondrial haplotype. In every individual, the mitochondrial DNA is exactly the same.

However, when Ricardo looked at lynxes from 2,000 years ago, from 5,000 years ago, from 11,000 years ago and from 50,000 years ago, every lynx still had the modern haplotype! Eureka! The clone-like uniformity of Iberian lynxes can't have been due to a massive population bottleneck in the twentieth century. Or even a massive population bottleneck at the end of the Pleistocene. Lynxes from 50,000 years ago – before modern humans had even made it to Spain – were identical too. Iberian lynxes had been living with diversity levels lower than inbred lab mice since before the height of the Ice Age. For Iberian lynxes, and hopefully Northern lynxes also, low genetic diversity is not the death sentence it was always thought to be.[*]

But it may all be a moot point. The biggest issue facing reintroduced lynxes is definitely not inbreeding. It's humans. Humans in cars, humans with guns, humans with snares and traps. There are now more than 300 Iberian lynxes in the Spanish wild thanks to intensive conservation management and care. In just one year, between 2014 and 2015, 21 lynxes were run over and killed on Spanish roads. Nearly 10 per cent of the population mown down in one typical year by drivers unwilling to arrive at their destination one second late. The more money invested in preserving Spanish lynxes, the more outrageous it is that those millions of euros are

---

[*] Some species of cat seem to be quite resilient to inbreeding. The Gir lions of India all have the same mitochondrial haplotype but seem to be thriving, and some African populations of cheetah are so genetically identical that you can graft skin between individuals without any issues of tissue rejection.

ending up twisted and broken on the baking tarmac. I can't even get mad: at least this is extinction by accident. As far as we know, no one is deliberately running them over.

In February 2000, 30 years after the reintroduction of the lynx to the Swiss Alps, the Bern environmental office received a heavy package through the post. Inside were the severed paws of a lynx, along with a postcard boldly proclaiming the gift as 'from the Bern hunting jungle'. A few days later a mother and her two cubs were found poisoned. This scenario could not be more emblematic of the real problems facing lynxes, native or reintroduced, if the parcel had been wrapped in a huge bow with a label saying 'metaphor alert' written in green pen. When the Swiss brought lynxes back to their country, there was no public consultation. There were no focus group meetings with farming community leaders or hunting groups. They were very much not onside. Lynxes prey mostly on deer and mid-sized wild ungulates, but the killing of domestic sheep is a recognised part of lynx behaviour – like Lilleth in Aberystwyth. Cattle and horses are too big to be part of the lynx diet, but sheep hit the sweet spot. How many sheep are lost to lynxes depends on many factors. Since lynxes are ambush predators, flocks grazing in open fields are safer than those in mixed woodland. Fenced enclosures are a very effective deterrent to lynxes too. In Scandinavia, as many as 15 sheep are lost per year for each lynx. In Slovakia it's more like four per year per lynx. Given the price of sheep at market, each living lynx represents a substantial cost, borne by farmers. Government compensation schemes can offset those costs but if the bureaucracy is too onerous, the simplest solution is to grab a gun. Luckily, in Switzerland, although the farmers and hunters took matters into their own paws, the general public had taken their lynxes to heart and were horrified by the slaughter. For the hunters and farmers, it was a spectacular own goal, serving only to paint themselves in a capricious and bloodthirsty light. Switzerland has also set up relocation programmes to move lynxes that have a predilection for sheep to areas with fewer opportunities for mutton gluttony. The compromise seems to have worked, and although poaching still occurs, nothing as sinister as the Bern parcel has happened again.

What does all this mean for lynxes in Britain? We lost them so long ago that they have passed entirely from memory and culture. We have no fairy stories about lynxes. Unlike with the wolf and bear, our perceptions haven't been coloured by tales of them eating grandma or frightening Goldilocks. We have to go back to a time before the country we recognise today even existed to find the smallest of hints as to how we lived with lynxes. One of the oldest poems in British literature is *Pais Dinogad*, written in Cumbric, the native tongue of north-western England. This seventh-century lullaby lionises the hunting prowess of a father, as a mother says to their son:

> Of all those that thy father reached with his lance,
> Wild boar, and lynx, and fox,
> None were spared that were not winged.

The identification of lynx relies on a translation of the Cumbric word for 'lynx', which is not accepted by all but seems solid to me. That this scrap of evidence only talks of lynx as a hunting trophy is all too realistic. That is pretty much it for written proof.

Reintroduction could work for lynxes in Britain, in a way that wouldn't be countenanced for bears and wolves. Britain is signed up to various European directives and international treaties that emphasise the need for member states to take steps to not only conserve the wildlife they have but to reintroduce the wildlife they have lost. Britain is signed up to the Bern Convention (1979) and the EC Habitats Directive (1992) that state as much.* Lynxes fit the bill. They have no special prejudices against them, and a lot of latent goodwill from the general cat-loving public. They are highly dependent on roe deer and to a lesser extent red deer populations that make up the vast majority of their diet. Britain has these in abundance. So much so that a significant budget is put towards culling them each year to prevent damage to the landscape. Britain, and especially Scotland, has plentiful connected woodland that is their natural habitat and is currently empty of wild carnivores. The danger to humans is nil. The danger to livestock is small but manageable. If any extinct carnivore is to be reintroduced, it must start with the lynx or not at all.

I would absolutely love to see lynxes back in the Highlands. Eating deer that are currently shot and left to rot on the hillsides. Think of the boost to tourism!

---

* I'm writing this in 2018, in the middle of Brexit chaos. Whether the current fad for short-sighted jingoism will lead to us withdrawing from anything containing the word 'European' remains to be seen.

People come from far and wide to look for reintroduced lynxes in the Swiss Alps and the Dinaric Alps. It would be the same for the British lynx. All our rare and localised wildlife has spurred the growth of cottage industries with guided tours, special boat trips, cuddly toys and pencil toppers. We can look to the white-tailed eagle for proof of that. This apex predator of the skies was hunted to extinction by 1918 in Britain, its last nests on the Isle of Skye. The white-tail is a spectacular creature, a flying barn door, significantly larger than the golden eagle, and one of the largest raptors on the planet. Despite concerns about it taking lambs, the white-tail was reintroduced to the Isle of Rum in 1975, and over a period of years, 82 young birds were taken from Norway and successfully released. The population has gone from strength to strength. The eagles have flown the nest and landed on the nearby islands of Mull and Skye. Stable, local nesting populations have now established all along the west coast of Scotland. Guided tours on land and boat are now a regular feature of local life. Twitchers coming the length and breadth of Britain contribute about £5 million annually to Mull's coffers, and support 110 jobs in what is an isolated, rural economy. Impacts on local farmers are negligible. The project has been such a success that further reintroductions of white-tailed eagles have now taken place in Wester Ross in the western Highlands, in eastern Scotland and in Killarney in Ireland.

This is the kind of template that should be used for lynx reintroductions to Britain. The Highlands of Scotland are a natural starting point. Experts have compared deer density and the connectedness of the landscape to estimate that the north-west of Scotland could contain a healthy population of up to 400 lynxes. Here there are deer, there are woods and there aren't people. Farming is a big part of the economy, especially sheep farming, but if the effort could be put in to win round the farming community, then it could work. Attractive compensation schemes for lost sheep – or even better, a Swedish system where farmers are paid an annual stipend just for having lynxes on their land, irrespective of losses – would do a lot to involve farmers in the success of the project.

Alternative sheep safeguarding methods, ranging from the sci-fi to the bizarre, have seriously been put forward. Fitting lynx GPS collars with radio-controlled anaesthetic syringes is an idea that could have been dreamt up by Philip K. Dick. Using GPS to constantly monitor the first generation of lynxes to be released, if they started to harry livestock, the anaesthetic could be triggered to knock the animal out. Habituating sheep flocks to a guard llama has also been put out there. The guard llama has the benefit of being low maintenance (you don't have to feed them, as they can graze with the sheep), and is less likely to pose problems to hikers and ramblers than guard dogs. If you've ever met a llama, you know how terrifying they can be. In the United States, guard llamas have been used to reduce sheep predation from coyotes by 90 per cent. That'll do, llama. That'll do.

I guess I'm passionate about lynx reintroduction because it seems so straightforward. When surveys have been done, 65 per cent of the British public are behind lynx reintroduction. In places where it has already been done, like Slovakia and Switzerland, farmers and rural communities have mostly supported the lynx, and been adequately compensated when problems have occurred. We have the space, we have the means and we have the support. I sincerely hope that lynxes come back and in a century or two will become such an integral part of Britain's wild spaces that no one will remember they were ever gone. It's not as far-fetched as you might think. It happened with red squirrels. Yes, our cuddly Squirrel Nutkins went extinct in Scotland in the eighteenth century. Populations of reds were reintroduced from England and Scandinavia. Today, they are rapidly giving way to the invasive American grey squirrel* but are holding out in Northumbria and the Highlands. Nobody remembers now that our beloved red-tailed rodent is actually the product of a reintroduction. When you do things right, people won't be sure you've done anything at all.

In preparation for writing this chapter and to get a feel for lynxes, I went to the Highland Wildlife Park at Kincraig in the Highlands, temporary home of Felicity the Puma, and now home to a rare captive population of Scottish wildcats and Eastern European lynxes. The lynxes in Kincraig would be ideal candidates for a reintroduction population. They are genetically diverse, and comfortable in a British climate. I spent ages peering through the chain-link trying to get a good view. The lynxes mostly slept on the top of their enclosure, except for one springy

---

* Grey squirrels were introduced to Britain by wealthy landowners who liked their novelty. The upper classes even gave them as gifts to each other. After escapes and releases, the greys have ousted the reds over much of the country.

kitten, only weeks old, that was bouncing all over its parents, eager to play. The marmalade-orange of their coats gleamed[*] in the Highland summer. One adult sloped around the enclosure, pacing back and forth on enormous padded feet. I wasn't the only one there. The lynx station was one of the most popular in the park, vying with tigers and polar bears for attention. I wanted to see them come back to what was their home. Not behind an electrified fence but free.

There is the public support for reintroduction. There are ways to mollify the agricultural interests that oppose them. Lynxes have been brought back to other European countries. They could be part of our native ecosystem again.

When I think back to my brief encounter with a wild Canada lynx and imagine that people could have the same experience in Scotland, my heart sings.

---

[*] 'Lynx' is related etymologically to the words 'gleam', 'light' and 'lucid', via the Latin root '*lux*' (and an earlier Proto-Indo-European root). In Western myth, lynxes had the best eyesight of all mammals, and the image of their powerfully shining eyes bound together the animal with the idea.

CHAPTER TEN

Grey Wolf

Extinct in Britain seventeenth/
eighteenth century AD

> From the moment the last wolf died, nature in the High-
> lands – in all Scotland, all Britain – lurched out of control.
> It still is out of control, and it will remain out of control
> until the day the wild wolf is put back.
>
> Jim Crumley, *The Last Wolf*, (2010)

The last wolf in Scotland had its throat slit at Ballachrochin
on the river Findhorn, east of Inverness, in 1743. A 'quine
and her twa bairns' had been travelling from Cawdor,
when a giant black wolf attacked. It dismembered the
children and stripped their flesh. The woman fled in
terror to Moy Hall, seat of Clan Mackintosh. Here, the
clan members were rallied, and signal was sent that a
great drive would begin the next day to bring the wolf
to bay and rid the land of its kind. While waiting at the
appointed time for all to gather, the laird grew impatient
waiting for MacQueen, his best ghillie. An hour after
everyone else, MacQueen deigned to show himself and
faced the wrath of the chief. 'What was the hurry?' said
MacQueen to the laird. Shaken by such insolence,
appeals were made for him to think of the poor, slain
children and the black menace that still stalked the
Findhorn Valley. With an air of theatrics that we
Highlanders are known for, MacQueen then drew back

his plaid, threw the wolf's head to the floor and cried, 'There it is!'

MacQueen: 2 metres (6.5 feet) tall, killed wolves by the hundreds using lightning from his eyes and thunderbolts from his arse. If you hadn't already guessed, there is nothing to substantiate this pervasive legend.

The actual last wolf in Scotland was killed by Sir Ewen Cameron of Lochiel, at Killiecrankie in the beautiful Garry Valley. Sir Ewen, dubbed the 'Ulysses of the Highlands', was a bear of a man who possessed prodigious strength and was able to draw blood from a simple handshake. Here, on the wooded banks, close to Soldier's Leap, he shot and killed Scotland's final wolf. Well, maybe. Sir Ewen was real, but whether he killed the last wolf or not is open to debate.

A man called Polson stabbed the real last wolf in Scotland around 1700, near Helmsdale in Sutherland. Out with his two boys, Polson came across a wolf den amongst the rocks. As the entrance was narrow, he could not enter. Instead, he sent in the boys, who, inured to the way of things, quickly killed the half-dozen cubs inside. Their feeble mewling alerted the nearby wolf mother, who streaked back to the den, breaking past Polson who had the wits to grab onto the animal's tail. Straining with all his might and shouting for the boys to leave quickly, he was asked, 'What is blocking out the light, so we can't see our way?'

The father, exhibiting immense callousness or incredible sangfroid, retorted: 'If the tail breaks, you will soon know.'

With the tail grasped firmly in one hand and the boys out of the den, Polson managed to reach his *sgian dubh* with the other and stab the wolf. Despite a memorial stone in a nondescript lay-by off the A9, there is as much substance to this story as the others.

I hope you will forgive me for crying wolf with these tales. It seemed appropriate.

The last wolves in Scotland were killed by giants with broadswords, fathers with dirks, cailleachs* with frying pans, soldiers with muskets, lairds and serfs, chiefs and peasants. According to the stories. The truth is, nobody knows when the last wolf lived. Nobody knows when the last wolf died. Nobody cared when it happened and nobody mourned when it was done.

In all probability, the last British wolf lived secretly and silently on the peripheries of human life and died of old age after failing to find a mate. Or it succumbed to disease after mating with any one of the thousands of feral dogs that had usurped the wolf's place in the woods and forests. There are no answers. Only legends.

Let's start with some facts. The wolf family comes from the Americas. Within the wolf family, there are also the foxes, the jackals and many species of wild dogs both well known and not. Like horses and camels, the Canidae (the taxonomic family that holds wolves

---

* One of my favourite 'last wolf' stories is of an old wifie travelling to Beauly from Struy in Strathglass to return a borrowed frying pan. By a strange coincidence, Struy is less than 15 kilometres (10 miles) from Cannich, temporary home of Felicity the Puma. Anyway, while taking a rest on her journey, she sat upon a stony cairn to catch her breath. Unbeknownst to her, the last wolf in Scotland had made these stones its den and thrust its head out when the woman sat down. Quick as a flash she brought the iron pan down on the wolf's head and killed it dead.

and dogs) have a New World beginning* and only later spread out into the Old World. The species formerly found in Britain and still found in Europe is the grey wolf (*Canis lupus*). Its descent can be traced back to the fossil Mosbach wolf (*Canis mosbachensis*) from the Middle Pleistocene. Its fossil record in Britain is pretty strong. There is a good selection from Oreston Cave in Devon,† and a particularly fine Pleistocene wolf skull was unearthed by Reverend MacEnery from Kents Cavern (see also Chapter 3). Wolves of one sort or another have been in Britain for over a million years. They are a very successful group. You might even say that grey wolves had the greatest global range of any mammal before *Homo sapiens* came along.‡ At their peak they were found from Portugal to Mexico and in every habitat except tropical forest and deep desert. For such an adaptable and catholic carnivore, it's amazing we've managed to exterminate so many of them.

The wolf holds a prominent place in the stories we tell ourselves; even the name 'wolf' has transferred itself onto species from other places and times. Dire wolves

---

* Today, the Americas have the greatest concentration of obscurely known canids including the bush dog, Darwin's fox and the crab-eating fox.

† The Oreston wolf bones are probably Late Pleistocene. Richard Owen (ugh) is on record as licking them – they stuck to his tongue. This is a useful test for distinguishing fossils old enough to have lost their organic content.

‡ Although you can quibble that lions may have had a greater range if you consider cave lions and American lions as part of the same species as modern lions.

are not merely an invention of George R. R. Martin but a real species that lived until the Late Pleistocene. We have more bones of dire wolves than nearly any other extinct Pleistocene animal, thanks to the wonderful preservation conditions in Rancho La Brea (see Chapter 3). Unlike their Westerosi namesake, real dire wolves were actually about the same size as a big husky but differed in proportions of the head and feet. They may be close relatives of grey wolves, or they could be on a completely separate branch of the family tree.

Red wolves (*Canis rufus*) are another kind of wolf, and one with an interesting backstory. They are a bit smaller than the grey wolf, and a bit bigger than the coyote. Found in the south-eastern states of the US, red wolves have been a conservation conundrum, hunted to near extinction in the first half of the twentieth century. By 1975 oblivion was imminent, and 14 individuals were captured to start a captive breeding population. They were just in time; shortly after, red wolves became extinct in the wild – the captive colony were the only survivors. Luckily, they bred well and have now been reintroduced to parts of Florida and North Carolina.

However, this straightforward success story has been seriously muddled by the advent of genetic testing. The first studies of red wolves were done to find out where they stood within the wolf family tree, but the results were unexpected. All the mitochondrial DNA signatures of red wolves matched either grey wolves or coyotes – both species whose range overlapped with the range of red wolves. Everything pointed to *Canis rufus* as not a distinct species but a hybrid between two closely related species, albeit one that was fully fertile. This is a huge problem for conservationists, because hybrids are afforded no legal protection, being neither one thing nor the other. The red wolf is now back in jeopardy, despite a healthy captive population. Humans, as a whole, are not very good with blurred boundaries, and that's exactly what the red wolf is. It's a hybrid that has carved out a niche for itself. Ironically, as a hybrid species ourselves, we've crafted laws that don't protect other hybrids. For the moment, red wolves are safe in captivity and in their reintroduced populations, while people argue about what to do with them.

The Falkland Islands wolf (*Dusicyon australis*) is a lupine enigma that even Darwin commented upon. When he visited the Falkland Islands in 1834,* it was the only native land mammal he encountered. It survived on nesting birds and the occasional seal or beached whale, and Darwin was puzzled about what kind of species it was and how it had arrived in the remote archipelago. Equally unusual was its tameness. The Falklands wolf neither feared humans, nor considered them prey. Such innocence was quickly taken advantage of. Darwin notes in his *The Voyage of the Beagle*: 'The Gauchos, also, have frequently killed them in the evening, by holding out a piece of meat in one hand, and in the other a knife to stick them.' At least four dead wolves made their way to Britain aboard the *Beagle*, perhaps killed in the efficient manner of the gauchos. Presciently, Darwin also wrote: 'Within a few years these islands shall have become regularly settled, [and] in all probability this [wolf] will

* Darwin stopped off at the islands twice, briefly, while voyaging on the *Beagle*. He wasn't impressed, saying they had 'an air of extreme desolation'.

be classed with the dodo, as an animal which has perished from the face of the earth.' And so it came to pass.

It's only recently that genetics have been able to answer the questions that perplexed Darwin. In spite of the extinction he predicted in the late nineteenth century, researchers have used ancient DNA extracted from one of the skulls collected by Darwin himself to look at their genes. The DNA shows that the Falklands wolf is a close relative of another extinct mainland South American canid: *Dusicyon avus*. Known only from fossils, it doesn't even have a common name, the binomial translating as 'ancestor of the foolish dog'. Nothing like kicking a dog when it's extinct. The genetic separation between *Dusicyon australis* and *Dusicyon avus* gives an idea of how the former reached the Falklands. Dating the split between the two lineages using the molecular clock[*] gives a divergence date in the Late Pleistocene, before humans arrived in Patagonia. People didn't take the Falkland Islands wolf to the Falklands – they got there themselves.

During the Pleistocene the sea level was lower when global water was locked up in glaciers on the land, and this exposed a vast area of continental shelf between Patagonia and the Falklands. The islands were never connected to the mainland, but it is likely that there was sea ice providing a land bridge for the animals to cross. As the ice melted and the sea rose again, the Falkland Islands wolf was isolated, left to hunt for stinking seals and mouldy penguins, surviving happily for 10,000 years

---

[*] The molecular clock is just a way of using an estimated rate of DNA mutation to calculate how long two branches in a phylogeny have been separating.

until just after the visit of an ambitious young English naturalist. Darwin gave them immortality in his writing while killing them for museum collections. It's such a senseless waste. With just a few decades' grace we might have recognised their uniqueness and saved them. They were discovered too early and we valued them too late. Now they are just one more extinction on an overcrowded list, museums their last remaining habitat.

These other wolves are a digression, though. The grey wolf is the only wolf in Western Europe. We think we know it: a merciless, wanton killer. Scourge of shepherds. Hunter of children. 'Watch out for false prophets. They come to you in sheep's clothing, but inwardly they are ferocious wolves.'* We have been deceived by false prophets. Our culture has become so saturated in wolf allegory, thanks to the Bible, Aesop's Fables and the Brothers Grimm, that the allegory has become the fact. At first, wolves were a cypher for the evil that men do. When everyone was a shepherd, the imagery of flocks, wolves, danger and safety was something everyone could instantly understand to make moral teachings intelligible. Now, instead of being the allegory for moral danger, wolves have come to be thought of as the danger itself. They have taken on the mantle of our wickedness and are persecuted everywhere, reduced in numbers across their entire range. Grey wolves are gone from most of France, Spain and Germany. They are extinct in Britain, extinct in Japan too. On the opposite edge of Eurasia, another archipelago had wolves and exterminated them like we did. Looking at scraps of bone preserved in

* Matthew 7:15.

museum collections, the DNA of Japanese wolves shows they were a unique lineage. Albeit one that was hunted to extinction by 1910. Nobody has looked at the DNA of British wolves yet.

The phylogeny of grey wolves has had decades of study and continues to throw up surprises. For whatever reason (massive bias, I reckon), there's been significantly more research performed on wolves than any species of cat. Still, the basics are clear. Grey wolves have a major split between populations in Eurasia and populations in North America. The split point is Beringia, that great, lost subcontinent that used to join Siberia and Alaska as one giant steppe. Beringia worked as a net exporter of grey wolves, a Xanadu full of meaty megafauna from where they continually expanded. All modern wolves can trace their origin to this region.* Modern wolves in northern Eurasia and in the Americas contain a vast number of genetic lineages, but surprisingly there doesn't appear to be any underlying geographic structure to their distribution. Unlike bears, which show the influence of retreat into refugia during the height of the Ice Age (see Chapter 8), wolves don't fit into neat refugial populations. Wolves are so adaptable that they didn't need to retire to Italy or the Balkans but could survive amongst the harsh steppe environment, hunting and scavenging on other megafauna. Wolves in India and the Himalayan region, however, do have their own unique

---

* And some extinct populations as well. Fossils, DNA and stable isotopes show that at the height of the Ice Age, Beringia had a hypercarnivorous population of grey wolves who appear to have exclusively hunted (and scavenged) the extinct megafauna.

and geographically isolated branches. These wolves form the most basal branches of the grey wolf family tree, and must have entered the Indian subcontinent early in the Late Pleistocene and been isolated from the rest of the species.

I've got to interject some personal viewpoints right about here. Well, it's my book. I like wolves, but I'm not a dog person. I'm not really a cat person either. I never had pets growing up and the appeal of them never really took root in me. I'm certain that part of the reason that so much time and money has been spent on wolf research is simply because people are interested in dogs. Dogs are wolves, plain and simple. Domestication altered the grey wolf into everything from the miniature Turnspit Dog* to the giant Great Dane.† Despite all the research on dogs and their relation to wolves, I'm not convinced that anything more than tentative conclusions have been reached. Precisely because wolves are large mobile hunters that don't seem to have been adversely affected by the Ice Age, it is very difficult to find any clear signals in their genetics that allow us to say anything concrete about the evolution of wolves or dogs.

We know that dogs were the first animals to be domesticated. When that occurred, though, is somewhat of a mystery. Most researchers accept a date of about 16,000 years ago. Some accept evidence of much earlier 'proto-dogs' even back to 36,000 years ago. It is a tricky

---

* A tiny extinct breed specifically bred for running on a wheel that turned the roasting spit over the kitchen fire in pre-industrial times.

† Actually a kind of mastiff originally bred in Germany.

and intractable problem. When does a wolf become a dog? Dogs are wolves, just a highly modified form. Teasing out of the archaeological record just when those modifications happened only tells us that domestication was going on long before then. DNA isn't much help either. Dogs and wolves form closely bunched genetic groups that say nothing about origin. Only a handful of Pleistocene wolves have been analysed genetically, and those confuse the picture further by showing variation that has gone extinct.

It's like trying to work out how Spanish and French relate to each other if all knowledge of Latin had been wiped out. The story is so convoluted that I had to phone up my friend Professor Greger Larson, who works at the University of Oxford, to try and make sense of some of the research. Greger is an expert in the genetics of dog domestication, and we've worked together on fun projects looking at pigs and chickens in the past. Greger is the polar opposite of me in many ways, not least because of his love of dogs. He currently runs the biggest collaboration of international researchers looking at the genetics of dog domestication. He grew up with dogs at home and they've clearly got under his skin. We chatted about some of the recent research on dog domestication, much of it produced by his group, and I tried to get a handle on what was going on. His interpretation seems to be one of a dynamic and reticulate domestication. Dogs are domesticated from an extinct population of wolves in Europe and also in Asia. That the two source populations of wolves have gone extinct is what has complicated the genetics of modern wolves and dogs so much, and shows the power of looking at ancient wolves to help understand the story. Mixing things up further, after agriculture had spread into Europe, dogs were transported from Asia and had the effect of diluting the signal of the original European dogs.

Today we are left with a situation where nearly all modern dogs can trace at least part of their ancestry to the Asian domestication. It's a fascinating story of domestication as a signal of human movements and human intentions in the past, when dogs were guards, hunting companions, muscle power and food store in one four-legged package.[*] The process has irreversibly changed the dog to meet human needs. And still, dogs are the world's favourite pet. Inside the famous Chauvet Cave, on the dried mud floor, there is a quiet reminder of how long ago the process started. In one of the chambers, the prints of a wolf, or a dog, follow along beside the footprints of a young child.[†] Here, more than 25,000 years ago, did someone walk with their pet? Or were they being stalked by a predator? It's the crucial dichotomy of wolf/dog that makes canines so difficult to study. When does the hated wolf become the beloved dog? What metrics of the skull allow us to separate the devourer of flocks from the playful puppy? There is no clear division. A dog is a wolf is a dog. It is something we would do well to remember.

[*] This is shown in the story of the Kurī, the extinct Māori dog. Taken on awesome island-hopping trips by Polynesian ancestors as part of their colonising package, the dogs were used for companionship, hunting, fur and, when necessary, food.
[†] Chauvet, despite its many masterpieces of prehistoric art and extinct animals, does not have a single painted image of a wolf. Quite extraordinary, when you think about it.

Tied up with the general fear of the wolf is the deliberate ignorance of how dangerous dogs are in comparison. There are many millions of dogs in close contact with us every day, compared to the paltry tens of thousands of wolves roaming the forests and plains. A conservative estimate gives a ratio of 5,000 dogs for each wolf left on this planet. Every week there are stories of dog bites, occasionally fatal. In England and Wales alone each year there are around 250,000 bites and maulings, with over 6,000 hospital admissions and between two and six fatalities from domestic dogs.

In stark contrast, wolves, where they coexist with people, always do less physical harm than dogs. Wolves do kill, there is no doubt about that, but it seems to be very much exacerbated by human poverty providing opportunity more than anything else. In Scandinavia, wolves have recovered from near extinction since the 1970s, and now at least a hundred animals range over Norway, Sweden and Finland. No attacks are recorded within the last century. In North America, no fatal predatory wolf attacks are recorded within the last century either. Predatory intent from wolves towards people in North America is so unusual that some detailed reports of 'aggressive behaviour' have been written up for scientific publication. In one 1985 paper by Scott *et al.*, the three authors out walking near Churchill, Manitoba, faced a surprise attack from three wolves. Written in a breathless style unusual for scientific literature, the report makes a thrilling read. After the initial surprise charge, the walkers used an air horn to startle the wolves and shinned up two trees. They stayed in the trees for four hours, watching as the wolves walked around them. Finally, when it seemed the wolves had moved off, the three climbed down and formed a back-to-back defensive triangle. With a 360-degree line of sight, the group walked from tree to tree back to their car. The encounter was unusual enough to warrant publication, despite not a single bite being received.

In Iran and India, where data has been tabulated, people live more intimately with wild wolves and the opportunities for conflict are consequently greater. In just one province of western Iran between 2001 and 2012, there were 53 attacks on people by wolves, and five of the victims died as a result of their injuries. In Bihar state, India, between 1993 and 1995, 80 attacks were reported, 60 of which were fatal. Looking at and breaking down the data gives more nuance on the threat of wolves. Fatal attacks are almost entirely on children under the age of 12. Left on their own while family members worked, they were vulnerable. The wolf could strike at defenceless youngsters playing in fields or walking unaccompanied near woods. Even within India and Iran, the attacks were deemed exceptional.

In normal situations the danger from wolves is close to zero, except when rabies is involved. Rabid wolves can be a particular problem, and are well documented for their tendency to travel dozens of miles biting anyone and everyone they can find. Although effective modern treatments for rabies exist, and the prevalence of rabies in the modern world is a fraction of what it used to be, fear of the disease is still strong – especially if you live in a region with poor health infrastructure. Western Europe has been free of endemic rabies for a century, but prior to this the disease terrified people. Infection with the rabies virus is by biting or scratching from an infected animal, with the virus transferred through the saliva. After a latent period, the virus multiplies and moves into the salivary glands, where it ensures that the saliva is full of infectious agent. Simultaneously, within the nervous system, rabies induces specific changes in the mind of the sufferer, all designed to aid the spread of the virus. When infection is full-blown, the dreaded hydrophobia becomes apparent. The sufferer can no longer swallow without excruciating pain and they show extreme aversion to even the sight of water. At the same time, they salivate constantly. As the rabies virus multiplies within the salivary glands, inability to drink prevents the viral load from being diluted, and constant salivation ensures that the new viruses can come in contact with more hosts. Alongside hydrophobia and hypersalivation, the victim can become overexcited and restless. Death follows soon after through cardiac arrest.

Today, feral dogs are the greatest source of rabies transmission. In the past it would have been wolves too. Wolves infected with rabies. Not a good combination at the best of times. I think it's extremely interesting that one of the strongest lupine archetypes we have in the West is that of the werewolf.* A bite from a cursed individual causes the newly infected to transform into an

---

* Rabies has also been implicated as the source for the vampire legend: biting, altered behaviour, irrational fears (holy water, garlic, crucifixes …). Everyone forgets that when Dracula first arrives in Whitby, he takes the form of a large dog. And bats are a reservoir of rabies as well.

animal with a deadly rage and an irrepressible desire to bite new victims. A clearer public service announcement of the epidemiology of rabies would be hard to craft.

The impact of wolves in Britain is as confused as in any other place. Our legends are of skulking brutes killed by heroes. Very little truth has come down to us from the past. Place names are one source of data. There's actually around about 200 towns and villages that derive their names from 'wolf'. Comically, one of the commonest wolf place names is 'Woolley'. In Devon and Dorset, Yorkshire and Berkshire, this name comes from the Old English for 'wolf's clearing' ('*wulfa leah*'). Likewise, 'Woolpit' in Essex and Suffolk (Old English '*wulf pytt*') is a place to trap and kill wolves. They are etymologic wolves in sheep's clothing. Additionally, there are way more places in Britain that have 'wolf' as part of their name but not due to the animal. It comes from the long history of 'Wolf' or 'Wulf' as a personal name. Many Anglo-Saxon kings and nobles took this as part of their name through fashion or happenstance. Æthelwulf, Wulfstan, Coenwulf, Wulfred and many more made their mark on British history. Wolverton and Wolverhampton are named after people, not animals.

I can remember when I first became aware that wolves had been a part of the British landscape and were not just confined to the distant Schwarzwald of Hansel and Gretel. In the Inverness Museum, past Felicity the Puma, there is an unparalleled collection of Pictish art, scooped up from various corners of the Highlands and dumped within the unlovely concrete walls of the museum. Amongst the treasures on display, one carving always stood out.

Map 10: UK sites mentioned in this chapter.

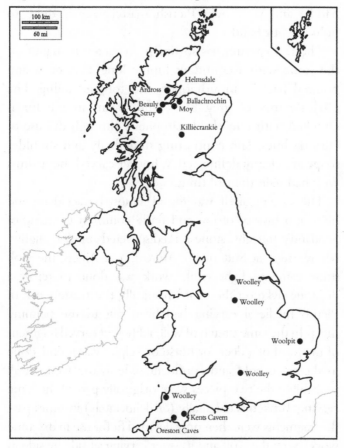

The Ardross wolf is a masterpiece. And not just to me. Judging by how it is used in the museum's promotional material, the Ardross wolf is a branding executive's dream. It is currently emblazoned on the side of the museum building, 9 metres (30 feet) high, visible throughout the centre of town. At the other end of the scale, you can get a squashed penny made with the Ardross wolf as a memento for only 50 pence. Put your coins in, wind the handle and your own Pictish masterpiece is there to hold in your hand.

The real carving has so much character. It strides off the stone with a confident lupine lope. Ears erect and pointed back – mouth open and tongue lolling. The intricate mix of grey, black and cream in the fur is chiselled with remarkable simplicity through the use of curving lines. The contrasting underbelly and shoulder stripe are clearly delineated. Whoever carved the Ardross wolf had seen the real thing.

The carving itself was found almost by accident, and hints at a history of use and reuse common to many of Scotland's ancient stones. Rediscovered in (or slightly before) 1891 at Stittenham, Ardross, near Alness, the wolf stone came to light while work was done to repair a drystone dyke on the estate. Ironically, this masterwork of Pictish medieval carving had been put to use penning sheep. In the same stretch of wall, a different carved fragment of the head of a deer (or horse or kelpie) was found, done in the same bold and confident style as the wolf. In all probability the two pieces were originally part of the same carving. Broken accidentally (or deliberately) in times past, the fragments were then only deemed fit for use in drystone work. Now, the wolf and its sister carving sit side by side in the Highland capital, reunited once more. The museum has even created a modern replica carving of the Ardross wolf. Directly opposite the original, visitors can touch the raspy sandstone and trace the outlines with their own fingers. It's here, in the presence of an image not so very old, that I, and hopefully others too, can feel a connection between the wolf and the place that used to be its home.

No issue in British conservation has been more contentious than the idea of reintroducing the wolf. Returning it to its ancestral home clashes strongly with powerful vested interests. Unlike the lynx, which would only affect livestock, the wolf poses a small but real risk to people too. In addition, no European government has enacted a successful wolf reintroduction programme. The return of wolves to Germany, France, northern Italy, Switzerland and Austria has happened organically. When human pressures were released, wolves were able to recover and naturally spread out to parts of their former range without political interference. People now having to share space with wolves are largely accepting of the change. It's amazing how much of a gulf there is emotionally between recolonisation by wild species versus reintroduction by humans. For this reason alone, seeing the return of the wolf to Britain seems a very long way off.

The story is very different in North America. The deliberate reintroduction of wolves to Yellowstone National Park in the 1990s still stands as one of the most audacious and wildly successful conservation interventions of all time. The wolves of Yellowstone restored much-needed balance to a degraded ecosystem. After the wolves' first extinction from the park in the 1930s, deer had been allowed to graze unchecked, completely destroying the previously balanced interactions between tree species like willow, aspen and cottonwood. In the absence of wolves, new trees failed to grow, and deer and coyote numbers exploded. When wolves were returned, deer were kept in check. More than this, the presence of wolves in the environment altered the behaviour of the deer, fear preventing them from grazing too long in one place, allowing the saplings time to regrow. Recovery of diverse tree types within Yellowstone provides new browse for moose, American bison and American beaver, promoting their recovery too. The simple act of returning a native carnivore produced rippling effects that spread out through the ecosystem, improving the diversity and tolerance of the whole.

The restoration of wolves to Yellowstone was a monumental act. Few other states have gone through the costly (both in terms of dollar value and political capital) process of wolf reintroduction. One example worth noting, for the way it sidestepped typically Western discussions about wolves, involves the state of Idaho. Although Yellowstone National Park does include a tiny finger of Idaho land (it pokes into Montana too), the potato state had its own role in bringing the wolf back. At the same time as the Yellowstone reintroduction, the federal government wanted to bring wolves to Idaho's northern Rocky Mountains as a potential bridge population between Canadian and Yellowstone packs. Unfortunately, the state government was far from keen. With persuasion from the powerful livestock industry and landowners in Idaho, the state legislature fought the federal government to such an extent that members of the Idaho Department of Fish and Game were legally prohibited from dealing with the federal wolf recovery programme. All attempts at compromise between state and federal authorities failed. Even the suggestion of delisting wolves from the Endangered Species Act to allow legal shooting of dangerous individuals didn't appease political opponents.

The issue of state rights trumped federal mandates. Wolves were one more casualty of realpolitik and the conflict between state and federal government.

Now, the Idaho House clearly didn't want to work with the federal government, but there was another state authority that did. The Nez Perce (Niimíipuu) tribal lands are in central Idaho. Where the state government had reached an impasse, the Nez Perce saw an opportunity. As the rightful inheritors of the land now called Idaho, the Nez Perce Tribal Executive Committee approved of partnership with the federal wolf recovery programme in the northern Rocky Mountains. A treaty between the two entities was written up. Translocated wolves would be released on Nez Perce land. Responsibility for monitoring the wolves, a difficult feat in the vast tracts of roadless land in central Idaho, would rest with the Nez Perce. If you've paid any attention to the outcome of treaties between the US government and Native American peoples, you know what's coming next. It shouldn't shock you to learn that the federal government, in the guise of the Fish and Wildlife Service, stiffed them on the bill, underfunding the expensive monitoring costs and leaving the Nez Perce to make up the shortfall through appeals to the Bureau of Indian Affairs and other grant bodies. Nonetheless, in January 1995 the first 15 Canadian wolves were released in central Idaho. Ten years later there were 450. Twenty years later there were 750. The Nez Perce, working in difficult conditions and with an underfunded budget, achieved what the state and federal governments could not: the return of the grey wolf to Idaho.

Could something similar ever be countenanced in Scotland? Given the successes in Idaho and Yellowstone, and the massive positive benefits wolves bring to deer-heavy ecosystems, you'd think that Scotland would be an ideal place to have wolves back. Red deer in Scotland are at artificially high numbers. The lack of predators has allowed them to reach densities that are unlike anywhere else. The UK government has also put in place targets for deer density. Numbers have to be kept down by culling (an expensive and difficult job) and paid hunts (exclusive and rare overall). In fact, stag trophy hunting has so fallen out of favour since the Victorian heyday that most shooting estates are not profit-making. At the same time, the mandated culling of deer hinds is a net loss as gamekeepers have to be paid and it's not worth the effort to process the animals to sell as venison. Wolves are an answer. When the effects are modelled, wolves could actually produce a net profit for estates just through their impact on deer. With wolves naturally reducing deer to sustainable numbers, the costs of culling are saved. Predation on sheep and cattle would have to be compensated for, as is done successfully in Scandinavia. Like for the lynx, ecotourism would provide a boost to the local economy too. Money talks, and though at the moment farmers have a more negative attitude to wolf reintroduction than either the rural or urban public, things can be turned around. It would take hard work, and there are no loopholes.* If we want the wolf back in Britain, we will have to convince the vested interests that

* A wolfy word. Loopholes were originally the thin defensive slits in castle walls allowing people to shoot arrows out but not let anything in – i.e. a loophole is something that can be exploited if you know how. The 'loop' in loophole comes from the French 'loup' for 'wolf', because those slits were your defence against wolves.

it will be to their benefit too. If this can't be achieved, then it will all be for naught.

Just last year, conservationists celebrated the return of wolves to Danish Jutland. A small pack from Germany had travelled more than 500 kilometres (300 miles) to settle in the region. It had been 200 years since Denmark had wild wolves. The only female was shot while walking through a field, by a man in a parked car. The whole thing was caught on video. It makes for distressing viewing. If we want to have wild wolves back in Britain, the visceral hatred that has been encouraged towards these animals must be turned around.

CHAPTER ELEVEN

Eurasian Beaver

Extinct in Britain sixteenth
century AD; returned
twenty-first century AD

> Cock up your beaver, and cock it fu' sprush,
> We'll over the border, and gie them a brush;
> There's somebody there we'll teach better behaviour,
> Hey, brave Johnie lad, cock up your beaver!

Robert Burns, 'Johnie Lad,
Cock Up Your Beaver' (1791)

Right. You can stop that sniggering. Though they've
been extinct here for half a millennium, beavers still
make a regular appearance in British slang. It's all very
well having a sly chortle to Burns's poem with its
amusing modern allusions, but have you ever wondered
why the beaver came to signify the pubic area? It's not a
straightforward story. People who have never seen wild
beaver (of any sort) may think that the euphemism is a
direct reference to the furry nature of both. That's not
the case. Beaver was originally slang for a man's beard.
The term reached peak popularity during the roaring
twenties when a schoolyard game imaginatively called
'Beaver!' was all the rage. The game consisted of shouting
'Beaver!' whenever you saw a beard. What fun! It's no
wonder that the same decade saw the invention of
television. Why beaver came to signify beards is not
certain. It could be because traded beaver pelts had a

passing resemblance to a beard. It could be because the part of a jousting helmet that protects the lower face is also sometimes called a beaver.* From the apex of beaver meaning beard in the 1920s, the term, um, drifted south; likely encouraged by the long-standing use of beard as a euphemism too. Of course, that's not the sense Burns is using it in his work, though. For most of the post-medieval period, a beaver was a hat (originally made of beaver, but later other materials were used). And not just any hat, but a hat as a symbol of wealth and prestige afforded by very few. Usually costing many months' wages for a working man or woman, they were exclusive to the moneyed classes. Johnie lad is fixing his hat (almost certainly not made of beaver fur), getting ready for a fight wi' the Sassenachs.

Beaver hats were expensive, and remarkable for a couple of reasons. Firstly, they are incredibly warm. The dense rich underfur of the beaver is impermeable to water and so thick it's practically solid.† Secondly, it is incredibly versatile and pliable as a material, and can be moulded by a milliner into any shape yet also retain it ever after too. Elegant top hats, Napoleonic bicornes, minutemen tricornes, wide-brimmed cavaliers and Puritan capotains were all made of beaver. The process involved felting the rich underfur, combing it out from the skins and processing it with heat or chemicals to

---

* Weirdly, nothing to do with the animal but from the French '*baviere*', or 'bib'. This is the sense used in *Hamlet* when Horatio says 'he wore his beaver up'.

† Beaver skin has somewhere between 12,000 and 23,000 hairs per square centimetre!

activate the individual hairs. Uniquely amongst furs, beaver hair has barbs along the shaft, an adaptation to keeping it waterproof and warm. The barbs are the key to the whole process, interlinking between hairs and making the felt malleable but stable – kind of like microscopic Stickle Bricks. The secret to felting used to rely on the fur being worn by native Canadians for a season or two. This form of pelt, known in the fur trade as '*castor gras*', was the most expensive beaver to trade. When worn as clothing, with hair on the inside, the coarse outer guard hairs were gradually rubbed away by the owner, whose sweat and body heat also worked the valuable underfur and activated the barbs. The second-hand coats could be sold direct to furriers for felting into hats.

Later on, chemical methods were developed to use on untreated beaver skins, known as '*castor sec*', bypassing the waiting period caused by having to first make coats for indigenous people to wear, and then, more importantly, get them back afterwards. The new treatment process used some nasty chemical salts, including mercury. Furriers and milliners working with beaver fur would invariably poison themselves from close contact with the ingredients and no understanding of health and safety. Mercury poisoning is interesting because it doesn't usually kill quickly but causes a vast range of neurological and psychological effects. Erratic mood swings, pathological shyness, muscle wasting in the arms, memory loss and formication* can all be present. Alice's

---

* The horrible feeling that you have bugs crawling around under your skin. I feel itchy just typing it.

Mad Hatter wasn't mad – he was slowly being poisoned by the tools of his trade.*

Appetite for beaver was unquenchable until the mid-twentieth century, when silk gradually took over as the fashionable and exclusive material for formal hats. Until then, the thirst for beaver fuelled not only the near extinction of our native European species but also a constant expansion into the Canadian wilds and Russian Far East, with trappers living off the money to be made from beaver pelts. The Canadian wilderness may have had more than 60 million beavers before Europeans turned up. Even by conservative estimates, at least 10 million of them ended up perched at jaunty angles on the heads of wealthy Europeans. Most of the rest spoiled before ever making it to market. What a waste.

The American wilds had to be opened up for beaver trade because European beaver populations had been destroyed. The American beaver (*Castor canadensis*) is a sister species to our Eurasian beaver (*Castor fiber*). The two look and behave essentially the same. For both species, the family is the core unit. Beavers are completely monogamous – a rare system in the animal kingdom – and the animals mate for life.† The entire colony usually consists of a mother and a father, their yearling kits and their newborns. The young stay with the group for the first year or so until their new baby siblings are old

* Not entirely a hazard of the past. Dental technicians working with making amalgams have sometimes reported symptoms of mercury poisoning through inhaling mercury vapour.
† Perhaps not as onerous a commitment as it seems when the lifespan of a beaver is only around a decade.

enough to take their place, and then are encouraged to leave and start out on their own.

These hard-working animals, synonymous with industry, will find a suitable new stream or pond to call their own and then, according to how deep and still the water is, modify it to make it home. Beavers' legendary ability to dam is a response to needing a still, deep area of water to build their lodge. The lodge is a defensive wooden castle surrounded by a moat of their own making. The entrance is underwater, a secret known only to the beavers, and the building is made from sticks and twigs trimmed from riverbank trees. With mud as mortar, the clever beasties make themselves a home to be proud of.

Beavers are strict vegetarians, feeding mainly on bark they plane from riverside trees. They are especially partial to nutritious young willow, ash, poplar and hazel. The large trees they cut down are not for use in dam building or other industry as they are too big to move. They are felled simply to get access to the high-up bark and the side-branches, which are used for dam building and food.* The mix of tree cuttings and rich silty mud in lodges and dams can even take root, organically growing a stronger structure. Beavers also stock larders for the winter. Collecting tasty twigs, they cache them underwater in their lodges. In harsh winters, when the

---

* As an example of how smart beavers are, the felling of large trees is not done randomly. Study in the field, looking at close to 1,000 beaver-cut trees, showed that nearly two-thirds of them were angled straight at the dam they were building, to make transport of wood easier. The little geniuses, working like tree surgeons, can make the wood fall where it will be easiest to move.

water freezes, they can still have access to food, safe in their insulated homes. Dams and lodges aren't the full extent of their building prowess, however. Like a Victorian industrialist, the monogamous beaver with its 2.4 children and banked wealth is also an expert canal builder. When trees are too far from the riverbank for comfort, using nothing but teeth and paws the beaver will excavate a watery channel so that wood can be floated where it is needed.

Water is safety to the beaver. They're one of the biggest members of the rodent family,* up to 30 kilograms (65 pounds) in weight. With no weapons, save their gnarly incisors, they can represent a sizeable meal to a predator. Like us. The meat is nondescript and slightly fatty – not something to get worked up about. However, beavers aren't like any other game animal. They spend most of their time between the land and the water – a liminal way of living. Sticking close to their pond on land, they use their naked tail as a warnihg device, slapping it hard when danger is about, to warn others. Broad and flat, the tail is scaly in a pattern that reminds people of fish.†
Scales like a fish ... lives sort of in the water like a fish. This was sound enough reasoning for many old-time Catholics to indulge in beaver during Lent, when red meat was usually forbidden. The 'semi-aquatic rodent is a fish' argument also got used for their cousin, the capybara, in Catholic South America. Beaver's not often on the menu today, but back at the height of fur trapping in the

---

* Only the South American capybara is bigger.
† Not really, but the tail does lack fur and has deep criss-crossing ridges, like the skin on your knuckles.

Americas, at least the meat could also be eaten throughout the year and sold along with the pelt. But that's not all.

Territorial as they are, beavers use scent-marking to stake out their home range. Adorably carting armfuls of mud from the river bottom, they make mounds around the perimeter and pat it down with humanlike hands.* Then they urinate all over the mounds, mixing the urine with a pungent pharmacopoeia of scented chemicals from specialist sacs that signal to other beavers that a territory is taken. Known as 'castor† sacs', the mix of secretions they produce is termed 'castoreum', and has been an exotic and luxury product for thousands of years. Extracted whole from the dead animals, the fresh sacs have a musky animal smell. Fur trappers, native and European, used to smear it on their traps to entice more inquisitive beavers into their snares. Crushed and boiled with alcohol, however, the subtle scents of castoreum are liberated and give notes of warm leather, vanilla and wood. Even today, it's used in high-end perfumes and as a food additive. Shalimar by Guerlain has always had notes of castoreum. You've probably eaten castoreum too. It was popular as a natural vanilla flavour in the past (a legal subtlety to differentiate it from actual natural

* The back paws are webbed for swimming, but the front paws have the five nimble fingers of a standard rodent.
† The Latin for beaver is '*fiber*' (as in the binomial name). Over time this has metamorphosed into the English word 'beaver' by an inversion of the *b* and *f* sounds. Many European languages share this inversion – e.g. the German and Yiddish for beaver is '*Bieber*', a name honouring those deemed industrious. The Greek for beaver is '*castor*' (also as in the binomial name), and this applies particularly to their scent production.

vanilla, and a good discussion point for those who naively equate natural with good and artificial with bad). Now, in the days of rampant orthorexia, flavouring your ice cream with bum scrapings from a dead beaver is not good for publicity, and very little castoreum is commercially used anymore.

In Roman times, castoreum was an essential part of the dispensary. Prescribed for all kinds of aches and pains, especially for women, it was seen as a panacea of sorts. It could be used as an abortifacient or an analgesic. Even up to the nineteenth century, castoreum was used as medicine by quacks and patent medicine sellers. It's true that thanks to the willow consumed in abundance by beavers, castoreum has a small percentage of salicylic acid.* It doesn't seem to be enough to produce any pain-relieving effects in humans, though. Wolves sometimes eat beavers too. In fact, research shows that wolves with lots of tapeworms will seek out beaver to eat. The usual thinking is that the change in diet prevents reinfection with tapeworms that pass through their usual prey of deer and other ungulates. I wonder if the low concentration of salicylic acid is enough to relieve the wolf's pain.

Before the demand for hats took off, beavers were mostly hunted for their castoreum. The Greeks mistook the castor sacs for testicles,† and later societies compounded the error. Fabulous stories are told about self-castrating beavers who would use their incisors to

---

* Willow bark (in Latin, *Salix*) has always been chewed for pain relief, as the bark contains salicylic acid. Aspirin (or acetylsalicylic acid) is a synthetic derivative of salicylic acid that avoids most of the stomach issues the natural acid can cause.

† The dummies. Beaver testicles are internal.

mutilate themselves and thus escape total destruction. How baby beavers were made, if their parents were so keen to bite their own balls off, is not clear from the tales. More comically still, it was claimed that twice-hunted beavers would stop to lift their legs to their pursuers, showing off that their valuable jewels had already gone.

Fur. Meat. Castoreum. Beavers were a hunter's boon – every piece sellable, and little danger in the catching. Even when the castoreum trade broke down in the post-Roman period, beavers were hunted for other reasons. In Britain, during Anglo-Saxon times in the sixth and seventh centuries, when beavers were still plentiful, people made necklaces from their incisors. About 16 graves have been excavated from this time period, where the occupants, exclusively young women or girls, had beaver tooth pendants. These weren't just any old knick-knack. Some had been capped with gold or bronze; precious yellow metals that would have complemented the bright orange* glow of fresh teeth. Why Anglo-Saxon women or their loved ones went to the expense of procuring beaver teeth (lower incisors preferred) and then crafting precious mounts for them is a total mystery. Perhaps the idea of beavers as a chaste animal, willing to happily sacrifice their gonads for their life, had filtered through from Greek and Roman stories. Beaver tooth pendants could have been the Anglo-Saxon equivalent of a purity ring, pushed onto girls as a way of controlling their sexuality before arranged marriage. At a time when

---

* If you've ever seen a live beaver, or even a photo, the teeth are practically tangerine. Beaver enamel is high in iron, which makes it tougher, hence the colour.

Christianity was replacing paganism, beavers could have told the same story as St Lucia.* Or perhaps they were something like the amber necklaces given to children today to supposedly help with teething. Beaver incisors are razor-sharp (they grind against each other to keep the edges fresh) and ever-growing (like in all rodents), which are some attributes that could have made them suitable as teething charms for young children or women of childbearing age. Alternatively, the suite of dental problems that sometimes accompany pregnancy may have been inspiration for them as a talisman of sorts. We'll probably never know, but the association between Anglo-Saxon women and beaver teeth is a strange one indeed.

An earlier association in Bronze Age England is more straightforward. Several graves from this time period have worked beaver bones interred in them. In particular, mandibles were neatly trimmed to remove the ascending ramus (the bit of bone that slots into the skull) and leave the flat bone of the mandible with only the incisor protruding. They are wood planes and chisels made of beaver mandibles. Hafted to a wooden handle by the bone, the tooth will happily cut chunks out of wood, just as it did when its hapless original owner was alive. The use of beaver jaws as woodworking tools is an incredible example of the ingenuity of people in a time before hard iron metalwork. Beaver planes seem to have an incredibly long pedigree, if we also take continental discoveries into account. Mesolithic sites in western Russia have hundreds of beaver mandibles that look to

---

* She plucked out her own eyes and gave them to a suitor who had admired them, rather than break her vow of chastity. It's why she is patron saint of the blind.

have been used in exactly the same way. Many even have tiny holes drilled through the incisor socket. Plugging the hole with a dowel would prevent the incisor moving when pressure was applied to the tool. Ingenious.

Actual evidence of their use in detailed woodcarving is scarce, since wood is so rarely preserved in the archaeological record. However, recent reanalysis of the enigmatic Shigir Idol, a towering wooden totem from the Mesolithic of western Russia, and the oldest wooden sculpture in the world, suggests that some of the finer work could have been done with beaver teeth. The idol, made from larch wood, has complicated geometric designs and faces all over it. Beaver planes, and even isolated incisors, would have been the right tool for the job.

Altogether, beavers were almost too good to be true – a one-stop shop combining hardware store, butcher, pharmacy and furrier. It's no wonder they were hunted mercilessly. In Europe, that meant that a century ago, only a few thousand were left at all, spread over eight relict populations: in the Elbe in Germany, the Rhône in France, Telemark in Norway and a few scattered river systems in Russia.

In Britain, beavers were probably gone from England and Wales by the fourteenth century* and from

---

* There is one tantalisingly late record of a bounty paid on a 'bever' from 1789 in the churchwarden's account for the village of Bolton Percy in Yorkshire. The history of this record is given in masterful detail in Bryony Coles's book *Beavers in Britain's Past*. Many researchers are sceptical of such a late survival, and it is possible that it might refer to a European river otter instead. However, bounties paid on otters in 1798 used the name 'otter' and were more expensive. Why would someone accept a lesser bounty for a 'bever' if it really was an otter? Why would the warden write 'bever' instead of 'otter'?

Map 11: UK sites mentioned in this chapter.

Scotland by the sixteenth century. They've left a rich legacy of place names. Beverley ('beaver's stream' in Old English) and Bardale Beck ('beaver's stream' in Old Norse) in Yorkshire are just some of the places that attest to living beavers in recent times. Written records have beavers sporting on the river Teifi in Wales. The saintly Gerald of Wales, archdeacon traveller and writer, describes either first-hand or close second-hand reports of beavers. He writes in 1188: 'The Teivi has another singular particularity, being the only river in Wales, or even in England, which has beavers; in Scotland they are said to be found in one river, but are very scarce.' It's the last secure record of an English or Welsh beaver.

The one Scottish river Gerald was talking about is probably the river Ness. The last written account of a Scottish beaver comes from Hector Boece, first rector of the University of Aberdeen. He writes in 1526 in his *Scotorum Historiae*: '*Ad Nessae lacus [...] Ad haec Marterilae, Fovinae ut vulgo, Vulpes, Mustellae, Fibri, Lutraeque in incomparabali numero, quorum tergoraa exterae gentes ad luxum immenso precio coemunt.*'* Strange that Felicity the Puma, a 'last' wolf and the last beavers should all come from this small patch of wild Highland country I call home. Boece's writing is the end of beavers in Britain, and his description of their use only as furs points towards the ultimate cause for their extirpation. Us.

Beaver fossils are relatively common in some parts of Britain. Predictably, they have been found in many middens, including at Edinburgh Castle. They have been found in peat bogs that were probably the end result of their own damming activities centuries ago. Some have been found in caves either as human detritus

---

* 'At Loch Ness [...] Here there are pine martens, foxes, weasels, *beavers*, otters in incomparable numbers, the luxurious furs of which foreign nations purchase at enormous price.' I know you didn't need me to translate this, but I'm doing it anyway for my own benefit.

or accidental entries. More evocative than the fossil bones, though, is the evidence of beavers found as trace fossils. Whole tree trunks that have been stripped of bark for food, or felled by beaver incisors, have been found perfectly preserved at some wet sites. Mesolithic Star Carr in Yorkshire contains the remains of a wooden platform at the edge of an old lake that was used by people, perhaps for fishing or preparing food. The trunks of birch have the characteristic pencil-sharpened ends of trees cut by beavers. Why go to the bother of chopping down trees with a blunt flint axe when you can get nature to do it for you? The wood at Star Carr may even have once been part of a beaver's lodge, repurposed by humans. Star Carr also has an abundance of beaver bones, some with cut marks. I would not be surprised if some of the bones were from the same beavers that had cut the trees. Beaver-gnawed trees have also been found at Westhay in Somerset, West Morriston Bog in Berwickshire and Flag Fen in Cambridgeshire, and you can even see a real example in the National Museum of Scotland. A poignant reminder that beavers were remodelling their habitat long before humans were.

Tree-felling and dam building are unique to today's beavers. Surprisingly, the roots of these behaviours do not go very deep into the geological past. One fossil species that did cut down trees is *Dipoides intermedius*. This small Pliocene (four to five million years old) beaver has been dug up from Ellesmere Island in the Canadian Arctic, an island which ironically enough has no trees today. The fossil wood, part of a nest or lodge made by the *Dipoides* beaver whose bones were still associated with it, was covered in tooth marks. Most of the extinct members of the beaver family did not engage in behaviours that we would call 'beaver-like'; however, they did come in a range of shapes and sizes.

In the Middle Pleistocene, Britain had a different kind of beaver inhabitant. *Trogontherium cuvieri* was a big boy. Half as large again as the Eurasian beaver of today, it was the ecological equivalent of a capybara. Its enormous incisors* couldn't do much woodcutting but were just right for clipping soft waterweeds. Munching on the lush waterside vegetation with prehensile Jagger lips, it lived when Britain was warm and tropical.

Larger still was *Castoroides ohioensis*, which disappeared at the end of the Ice Age. *Castoroides* gets a lot of attention because it was a giant. Like, the size of a bear. A bear-sized beaver! It lived in North America from Florida to Alaska during the Late Pleistocene, and overlapped with the first people to come in from Beringia. So far, we don't have any evidence of it being hunted, but given how we treated its little sibling, I wouldn't be surprised if it was a regular bag. As long as you stayed away from the incisors – those things would have been sharp enough to take an arm off, easy. Imagine how much wood they could plane! *Castoroides* is actually a pretty distant relative of *Castor*. There's no fossil evidence of it cutting down trees or making dams and lodges, though. It likely ate

---

* *Trogontherium* incisors could be up to 18 centimetres (7 inches) long, including the root within the mandible. That's the same size as the canine of a sabretooth cat.

succulent water plants and grasses like a giant rodent hippo. Similar to how the capybara lives today but much bigger. Detailed analysis of proteins from fossil *Castoroides* puts it in the rodent family tree, a long way from other beavers.

People have used genetics to look at Eurasian beavers, and even extinct British beavers. Scientists at the Natural History Museum in London took small fragments of DNA from fossil beavers from Burwell Fen and Gough's Cave, and compared them to modern and ancient European beavers. Reassuringly, beavers have a very simple genealogical tree. All populations can be placed into a western group or an eastern group[*] based on which side they fall of an imaginary vertical line drawn straight through the middle of Poland and extending north and south. British beavers belong with the western group, alongside the survivors from the Rhône, Elbe and Telemark river systems.

It's just as well, because the Telemark beavers were the source of Britain's first legal beaver reintroduction back in 2009. On 28 and 29 May of that year, 11 Norwegian beavers were released, with full governmental approval, onto three lochs in the Knapdale Forest region of Argyll in south-west Scotland. Frid and Frank, Katrina and Bjornar, Gunna Rita and Andreas Bjorn, and all their kits, were furry pioneers let free to roam Loch Linnhe, Loch Coille-Bharr and Loch Creag Mhor as part of an ambitious project to bring beavers back to Britain. Beavers are so inoffensive, so industrious and so

---

[*] With one exception: the extinct population of beavers from the Danube river system are a unique branch of the beaver tree.

charismatic that they had many champions fighting for their return. Through the work of the Royal Zoological Society of Scotland and the Scottish Wildlife Trust, working with the Forestry Commission Scotland and Scottish Natural Heritage, an application was put forward based on EU Council Directive EC/92/43/ EEC. The directive plainly says that all member states should

> study the desirability of reintroducing species in Annex IV that are native to their territory where this might contribute to their conservation, provided that an investigation, also taking into account experience in other member states or elsewhere, has established that such reintroduction contributes effectively to re-establishing these species at a favourable conservation status and that it takes place only after proper consultation of the public concerned.

After extensive local and expert consultation and a presentation of detailed plans, the government was convinced and the Scottish Beaver Trial began. In Knapdale the original 11 animals, supplemented with another five Norwegians* the following year, have gone from strength to strength. Closely monitored by an army of professionals and volunteers for their effects on woodland, water levels, plant diversity and their influence on amphibians, reptiles, birds and other mammals, they've become a cottage industry unto themselves. Thankfully, negative impacts are negligible

---

* Elaine, Eoghann, Christian, Trude and Tallak. A fine mix of Scandi-Celt names all.

and the positive impacts are almost too numerous to mention. Like us, beavers are ecosystem engineers par excellence, and their industrious felling and building modify nature in a way that nothing else can. The downed trees produce a coppicing effect, letting sunlight down to the woodland floor and encouraging new growth and new communities of plants to thrive. Their damming makes new water microclimates, giving insects like dragonflies and water beetles fresh hunting grounds. Standing trees, waterlogged by dams and dead on their feet, give new homes to woodpeckers and the grubs they feed on. Deeper, stiller ponds are a refuge to freshwater fish, who can enjoy the increased number and variety of invertebrates as food.

One particular worry, given how much they contribute to our shared culture and, more importantly, the Scottish economy, was what effect beaver dams would have on our famous diadromous fish, the Atlantic salmon. Salmon return home from the sea to spawn, and their spectacular upstream migration to spawning grounds relies on a lack of dams – either man- or beaver-made. In other countries with major salmon fisheries and beavers, for example Norway, the effect of beavers has been investigated. It's important to remember that salmon and beavers lived in Scotland (and everywhere else in Europe) quite happily together long before humans had even left Africa. It would be strange indeed if species that had lived in harmony in the past were to somehow become antagonists in the present. The reality is that salmon usually have no problem navigating past beaver dams, which tend to be no more than a few feet high – and leaky to boot. This is a happy contrast to the concrete behemoths we have placed across many rivers, though at least they now come with accessible salmon ladders. Beaver dams can increase siltation of the gravelly riverbeds that salmon use to lay their eggs, but the effect is likely negligible. It's more than offset by the proven benefits of beaver ponds and their rich bug life, leading to increased size of trout and other residents.

Close monitoring of the Knapdale beavers of the Scottish Beaver Trial lasted for five years between 2009 and 2014. A year after their introduction, the first kit, Barney,* was born to Frid and Frank on Loch Linnhe. When last surveyed, the population was stable. No increase but no decrease either. Most of the original animals released are dead but their kits have replaced them. They are now self-sustaining and loved by the local people. A population of beavers established in west Scotland for the first time in 500 years, legally mandated and protected. Legal beavers!

Now, the mere fact that there are legal beavers suggests that there must also be a corollary. And indeed, that is very much the case. While the government was crossing the t's and dotting the lower-case j's on the Knapdale experiment, illegal beavers were spotted in the east of Scotland. And they were legion. On the silvery Tay, the river Almond, river Earn, river Isla and river Tummel, as well as associated lochs and lochans, they had made their home. Canny Taysiders had kept

---

* Barney, and a later kit, Woody, were named by local school-children.

schtum about the illicit beavers in their midst, which
had supposedly been present since at least 2006.* No
one knows how beavers got to Tayside. No one knows
who released them. No one knows how many founders
there were.

It was a massive headache for the Scottish
government. While their flagship reintroduction was
happening in the west, a more successful reintroduction
had already happened, opaquely, in the east. The initial
plan was to round up the Tayside beavers into
quarantine for testing to make sure they weren't
infested with damaging parasites or potential diseases
(as had already been done for the Knapdale beavers).
With field surveys, the naivety of this plan was
exposed. Volunteers found 25 lodges and 39 territories,
with somewhere north of 150 beavers. Too many to
capture; too many to cull. Scottish Natural Heritage
instead managed to enfold the Tayside beavers into a
larger Scottish Beaver Trial, incumbent on detailed
study of the population to ensure that they were
compatible with the goals of the larger project.
A first priority was genetic testing. Where did the
Tayside beavers come from? Thankfully, DNA clearly
showed that they were Eurasian beavers rather than
American beavers and belonged to the western branch
of the family tree, likely from a mixed population in
Bavaria.

Differentiating between the American and Eurasian
beaver species is not a trivial matter. They look identical

* Perhaps even as early as 2001; a canoeist supposedly saw a beaver
on the river Earn on this date.

even to the trained eye, but under the skin there are some crucial differences. For a start, the European beaver has eight fewer chromosomes than its American sister. Hybrids are impossible, and they've been evolving apart for at least seven million years – that's more time than the separation of humans from chimpanzees. In more carefree, less bureaucratic times,* seven American beavers were introduced to Finland under the assumption that one beaver was as good as another. Because of this lax attitude, 12,500 American beavers can be found in Karelia (the region between Finland and Russia) today, taking up space that should by rights be occupied by native Eurasian beavers.

In a similar vein, during the 1870s, the Third Marquess of Bute decided that he wanted some beavers, and damnitall, he should have them! Since he owned the Isle of Bute, nobody could stop him releasing 11 American beavers onto his estate. At first, they did quite well. Perhaps up to 16 lived there during the 1880s. Somewhat inevitably, by 1890 they had gone extinct again, like the Marquess himself.

The fundamental issue is that while Eurasian beavers are native and good and true, American beavers are exotic and unwelcome. We want the beaver back but it has to be the right species. In the interplay between environment and animal, beavers basically extensively remodel their habitat to their liking. In Britain, the environment has had aeons to respond to this constant attack and fight back. A truce of sorts has formed, and

---

* 1935. They had moxie back then.

the beavers' ecological jostling does not perturb the new balance.

It is the height of irresponsibility to introduce non-native species to areas where they have never lived. Unintended consequences always come from even small species.* With a species that engineers its environment, like the beaver, releasing them where they have never existed can cause utter devastation. Sadly, this is exactly what has happened in the southern reaches of Tierra del Fuego in distant Patagonia. Here at the tip of Cape Horn, 25 American beavers were released in 1946, with the stated aim of providing a new exploitable resource for those that lived there. Terrible idea. Patagonia has never had beavers, they have no native predators there and the trees have not evolved defences to their presence. Beavers have felled vast stands of southern beech in their new home, spreading to other islets and creating dead bogland where it has no right to be. The long-reaching effects of beavers on Fuegian ecosystems is only just being quantified, but suffice to say, it's not looking good. Beavers don't belong there, and it's only because people thought they could make a quick buck hunting them that they are there at all. American beavers belong in North America. Eurasian beavers belong in Eurasia. Part of the reason why the Scottish Beaver Trial has taken so many pains to understand all the variables involved with bringing beavers back is because of the

---

* I'm thinking here of the utter decimation of wildlife caused by cane toads and rabbits in Australia, or by possums and cats in New Zealand.

potential for ecosystem harm. Despite being native, much has changed in the 400 years they weren't around, and once they are out of their transport cages it's very hard to get them back in.

The Scottish trial absorbs most of the press about reintroducing beavers to Britain, but England and Wales have also got in on the action. Like Scotland, England and Wales have a mix of legal and illegal beavers. Who knew that our small island was teeming with so many unlicensed beavers? It seems it's only when people started looking for them that they began to already find them everywhere. Legal, fenced releases have been conducted in Gloucestershire (Forest of Dean, Flagham Fen*), Kent (Ham Fen) and Cornwall (Ladock). Amusingly enough, just like in Scotland, the largest population of free-ranging beavers in England are from an illegal release. The river Otter in Devon has a few dozen beavers making their home there, original source unknown.

With thoughts of British beavers sluicing around my mind, I resolved that I had to at least try and see some in the wild. The river Beauly near Inverness has been embroiled in beaver controversy since the

---

* The Flagham Fen beavers were released on private land owned by the property millionaire Jeremy Paxton in 2005. The animals produced the first recorded native-born English beavers since their extinction in the fourteenth century. The six original releases were named Tony, Cherie, Gordon, Sarah, John and Pauline. I can't quite get my head around whether naming individuals of the world's second-largest rodent after the top tier of the UK Labour Party and their wives was intended as a compliment or an insult.

discovery of illegal free-roaming beavers there a few years back. They were living quite happily and bothering no one, but the powers that be decided they should be quarantined for study and possible relocation. In the process of trapping them, they died. Entirely coincidentally, and I'm sure in no way related to the mysterious appearance of those beavers, Aigas Field Centre beside the river Beauly also has a wee lochan within its fenced boundaries that has been stocked with a family of Bavarian beavers since 2006. I made arrangements with the estate, owned by naturalist Sir John Lister-Kaye, to go and visit for a few hours at dusk to try and spot the famous inhabitants.

It was a strange day, all told. In the morning I visited Ardersier, a small village on the other side of the Moray Firth from Beauly, to help my extended family put up a memorial bench to my grandparents. They were kind, hard-working and much-loved people who died within just a few months of each other after 66 years of marriage. Laughing and reminiscing with my cousins, we made good progress on concreting the foundations for the stone seat. It looked out across the firth where my grandparents had taken multiple generations of our family for picnics and holidays. Passers-by stopped to ask what we were doing and we took the time to tell them about who we were commemorating and their connection to the area. One old man even told us that he had had a dream the night before that a bench was to be put in the exact spot where we were working. I'm no believer in signs and omens, second sight or *taibh-searachd*, but I found this old Highland bodach to be very reassuring. It made sense to me in that moment, while recognising my connection to the personal past, to go and look for beavers.

I left Inverness at quarter past seven after a light dinner and drove the winding road to Aigas. At the estate, I quickly made my way to a hide beside the loch, binoculars and camera strapped to me, in the hope of making a connection to our shared natural past. I was the only person there. Gulping down the fresh scent of pine sap breezing in through the hide windows I sat down and looked. There was the lodge: a castle of silver sticks beside the far bank, humpbacking out of the mirrored water. Silence was everywhere. You forget what silence feels like in the hustle of everyday life. Here, it was like a warm cloak, wrapping itself around you, the quiet only broken by the occasional *hoo-hoo* of a woodpigeon in the trees. The lochan itself was beautiful: dark and peaty, with a patchwork of lilies on the surface and reeds at the edges. It looked uncannily right to my eyes. The way it should be. Somehow better and more beautiful than the glorified puddles that many lochans manage to be. I settled in for the hour of dusk, sweeping the view with my binoculars for any sign of the elusive beavers. Every ripple of the surface caused a surge of adrenaline. Every fish that breached demanded my full attention. It was exciting sitting down in the silence. I felt like I was on a quest.

There were no beavers roaming in the gloaming. As the light and my hope faded, I remembered the first rule of biology: 'nothing behaves the way you want it to'. All I had seen were some ruddy ducks – I think they were shovelers. Walking back to the car, begrudgingly, in the dark, and with many stops to turn and gaze wistfully at the water, I tried to evaluate what I was feeling. I won't deny that it was a disappointment not to see any beavers. But, for me, it was enough just to see what they have done. How they have improved the land just by being there, with their elegant lodges and constant landscaping. I had experienced an hour of perfect mental clarity, able to filter out the petty annoyances of everyday life to concentrate every neuron towards a simple goal. I hadn't seen a beaver but in a sense it didn't matter. I knew I'd be back again. You pays your money and you takes your chances. I and many other nature lovers know that the quiet searching is often as good as the seeing.

The beaver is different from the wolf, the lynx and the bear. Of all the mammal inhabitants of Britain, it is the only species that has ever been brought back. Think about this for a second. Turn on the news. Read the papers. Surf the Web. Whenever there is a story about the natural world, it is invariably bad. The signs of the times are poaching, trophy hunting, destruction and extinction. For once, just for once, there is a sliver of good to talk about. An old wrong has been set right. We did the honourable thing for once. It should make you feel proud. It makes me feel proud. In a world of tragedy and hurt, this success needs to be bellowed in the faces of everyone. We can fix things. It's not always too late. The return of beavers signifies the successful completion of tens of thousands of hours of work by hundreds of committed individuals, pooling resources towards a shared dream. It is nothing less than a miracle – the second coming of beavers. After an absence of 500 years, anyone can now go and see beavers born in Britain, sporting on any number of rivers. They are back for good.

CHAPTER TWELVE
The Future

> Consider the auk;
> Becoming extinct because he forgot how to fly, and could
> only walk.
> Consider man, who may well become extinct
> Because he forgot how to walk and learned how to fly
> before he thought.
>
> Ogden Nash, 'A Caution to Everybody'

Well, there you have it. If I've been at all competent in my crusade, you should now have the ability to see ghosts around every corner. Did a mammoth once stand here? Did a sabretooth once hunt there? I would love you to close this book thinking about the spaces that extinction has left unfilled, in Britain and everywhere else. We are all walking in a world of shadows, whether we recognise it or not.

Our planet has changed immeasurably over the course of the Pleistocene and Holocene, and what is left is a drabber world, pruned of all the strangest, largest and wildest animals. My strongest contention is that the ultimate cause for this loss is our own species, *Homo sapiens sapiens*. Spreading out from Africa, wherever we met megafauna they disappeared into a black hole.* In some places – the Americas and Australasia – the extinction was sudden and synchronised, taking a few

---

* The black hole between nose and chin.

thousand years at most after first contact. Many of the giant species that humans first encountered must have been naive, like the doomed Steller's sea cow. Unable to recognise the unique threat that humans posed, and unable to mount an effective response, they would have been easy to kill. Human hunting skills, fashioned from instinct and honed by culture, make them deadlier than anything evolution has produced before. If some species were adaptable enough to mount a defence with tusk and claw, it wasn't enough. It didn't take much to push those same species over the edge. They were slow-reproducing giants, investing much of their energy in one or two offspring every few years and relying on enough of them surviving to maturity to continue the species. Overkill was not a wanton slaughter but a slow (to human eyes) war of attrition and constant downward slide, year upon year of taking slightly more than could be replaced. Humans, in their supreme adaptability, can switch between prey species so that when any of their choices are hunted to exhaustion, human population numbers don't drop. Any attempt at population recovery in prey becomes impossible under the relentless pressure of hungry humans.

In Europe the simple story is complicated by the presence of our cousins, the Neanderthals. European mammoths, rhinos, shelk and the other megafauna were not naive to the threat that small bipeds could pose. Neanderthals had been hunting them for hundreds of thousands of years with no detectable extinctions. In Europe the extinctions are more staggered, possibly because of the wariness of the species to the threat they faced. Neanderthals themselves were one of the first megafaunal species to vanish from Europe under the onslaught of our own species.* This may have been due to simple competition for resources. But it might not. There are hints of darker things going on. Cut marks on the jawbones of a Neanderthal child have been found associated with modern human cultural refuse at Les Rois in south-west France. The cut marks are just like those on more traditional food species found at the same site. The implications are blackly clear.† The entire history of contact between human groups in the modern period shows what happens when there is any disparity in group power. Ask the native Tasmanians.

After the disappearance of Neanderthals, and some other species (cave hyaenas, cave bears, perhaps sabretooths too), things pause for a while. All were at the top of the food chain, or highly specialised, meaning low population sizes would be likely, making it even easier for them to be pushed over the edge. Although the climate was cold and dry, modern humans spread out everywhere north during this time and populations started to steadily increase.

That population increase is key to the later extinctions. At that point in time, humans had incredibly sophisticated stone and ivory weapons, widespread cultures (think of

---

* Almost. Remember, there is still some Neanderthal left in non-Africans, buried in their DNA.

† I personally find it very easy to believe that people could have treated Neanderthals as a food source as well as potential mates. Many of the Neanderthal bones that have survived show signs of being defleshed. Other Neanderthals, either for cultural or culinary reasons, could have done this; modern humans could have done it too.

the *Löwenmensch* and Venus statues) and long-distance
trade links for passing ideas and materials. The increase
in population size and sophistication of weapons is what
gradually doomed the megamammals that were hunted
to fuel human expansion. Pushed into ever-more
marginal territories to the north and west, Beringia and
Siberia were their last refuges, as well as the few islands
(either geographic or ecological) that could support
them. Mammoth, rhino and shelk populations all start to
collapse around 14,000 years ago, when Eurasian human
populations are at a high plateau. The lonely survivors –
shelk in the Urals and mammoths on Wrangel Island –
eke out an existence until even their refugia are violated
by humans.

Once the last of the megamammals are gone, at the
start of the Holocene, that's when things are irreversibly
changed. Without the woolly mammoths and woolly
rhinos to churn up the mammoth steppe, fertilise it with
their dung or keep it productive, the land itself begins
to change. It begins to turn from steppe to tundra and
taiga. The entire ecosystem that flowed from copious
mammoth dung died away completely. The medium-
sized mammals that could replenish their numbers better
than the megamammals suffered, both from the loss of
their optimum habitat and increased human hunting
pressure as they switched to whatever prey species could
be found. Some went extinct; others sought protection
in the forests. Carnivores like the cave lion now came
into more direct (and indirect) conflict with humans.
There was only one possible conclusion. And so, the
sixth extinction continued. And accelerated along with
our sophistication, rolling up the aurochs, pruning away

the moose, reindeer, brown bear, wolf, lynx and beaver. That's how I think it happened.

There are of course many people who take great issue with the idea that our own species could have decimated such enormous swathes of the planet. Climate change is put forth as the causal agent, with no (or minor) input from humans. How could we have killed them all with pointy sticks? We could, and we did. Climate was naturally changing in the background, following the Milankovitch cycles like clockwork, but there is no mechanism for the selective, staggered extinctions we see. Climate proponents are fond of saying that the overkill argument is circular, or that correlation doesn't equal causation. I think they are often guilty of this themselves. One of the arguments used against human overkill in the Americas is that for some species there are no radiocarbon dates in the timeframe that humans are around, so they must have gone extinct early, so it couldn't have been humans. This is just lack of data masquerading as a hypothesis.* For every extinct American species where there are more than a handful of radiocarbon dates, overlap with humans is shown.

That's not to say there aren't problems that human overkill struggles to fully explain. American mastodons seem to disappear from Alaska before humans arrive, maybe as early as 75,000 years ago. We have a good number of radiocarbon dates for this species from this

---

* Some might say the same about the scanty record of archaeological sites with megafauna. But there are clear signs of hunting in the archaeological record. For climate, you would expect many examples of species extinction before human arrival.

area, so it is an anomaly. Mastodons in the southern United States definitely overlapped with humans and have been found with hunting artefacts (the Manis mastodon, for one). Perhaps later Alaskan mastodons or earlier Alaskan humans will one day be found. Or this could have been one of the marginal populations most vulnerable to the natural climate change of the Pleistocene. The challenging switches between warm and cold would have negatively affected many species – it's just that humans gave the final blow.

Another example that I've scratched my head over is the Lord Howe Island horned tortoise. Why would a giant armoured land tortoise have gone extinct on this small uninhabited island, undiscovered until the eighteenth century? It had relatives that survived into the mid-Holocene in Vanuatu, Fiji and New Caledonia and they were definitely eaten by people on some islands. Perhaps the Lord Howe species was just one of the background extinctions. It's not a very convincing argument. There are also a few small and medium-sized species that would have had little appeal to human hunters but that have still disappeared. One must conjure potentially unknowable ecological interactions with either extinct species or habitat to explain them away. It's also difficult to quantify the additional pressures that deliberate human landscape burning and modification (e.g. in Australia and the Americas) and domestic or feral dogs (e.g. in northern Eurasia and the Americas) could have had on extinct Pleistocene species. Nonetheless, I hope you will agree with me that the buck must stop with humans.

What to do, what to do? Is it too late?

Now is the time to inject a bit of optimism. Talented and dedicated people are working on trying to claw us back some biodiversity. Going back to basics means learning a new set of the three Rs: reintroduction, rewilding and resurrection.

We have already met reintroduction. It is just transporting members of the same species back into their former habitats. For British mammals, the beaver is our single attempt so far. As absolutely adorable and beneficial as I, and many others, find them, they are not universally loved. They do have an impact on the land where they live. It's an impact that the land evolved with and brings many benefits, but it is an impact all the same. A minority of farmers and landowners can be loath to tolerate any damage to what they perceive as 'theirs'. Beavers are doubly damned because of their current nebulous legal status in Britain. Because many rivers have beavers that were illegally released, their protection isn't guaranteed. The Tayside beavers have suffered most under this twilight status. Twenty-one of them have been shot since 2010, including pregnant and nursing females. While negotiations were going on to bring legal protection, anecdotal evidence says that some farmers and landowners increased their lethal persecution, trying to get rid of as many as possible before the arrival of legal consequences. The Scottish Executive has moved pretty quickly and as of 2019, beavers in Tayside and Argyll are protected under EU directives. English beavers are still in the grey zone. And even with the legal protections provided under the EU, landowners are still allowed to use lethal force with 'problem' animals.

But these are all basically teething issues. There is a lot of goodwill from the public and from the majority of farmers and landowners that live with beavers. Where issues arise, it's mostly because there are no natural predators to keep populations vigilant and numbers in check.

Beavers are proof that British people are willing to live with some reintroduced species. The next stage, and a real test of how serious we take conservation in the UK, will be whether we ever see the return of any of our missing predators. From bears, wolves and lynxes, it's the cat that holds most promise. Ecologists have been, and are, working tirelessly to sell the reintroduction of this magnificent creature to the public. At the time of writing, various groups have put forth proposals to start trial reintroductions in England and Scotland. Time, and politics, will show how successful they will be. I would absolutely love to see lynxes back in the Highlands, or Northumberland, or literally anywhere. They are quiet, secretive, magical creatures who pose no threat to humans. If reintroduction is done right, they will have little impact on livestock and a massive positive effect on deer management. They could save us all money wasted on culling deer.

Despite all the ink that's been splashed on wolf reintroduction, I don't see it happening before lynxes are back. It could work but the potential for it to go wrong is too high. Without a previous successful carnivore project under the belt to point to and iron out the intricacies, it would feel dangerous.

Bears are another matter entirely. No one has seriously considered bringing them back to Britain, but why not?

Reintroductions have been tried in Poland, France, Austria and Italy, though on an extremely small scale. The movement of fewer than a dozen bears into Poland did not result in a viable population. Seeding of extremely tiny (fewer than 10 individuals) surviving populations of bears in the Pyrenees, the southern Alps and the eastern Alps, with two to three new bears in each case, has had mixed success. Until conservationists have better results with them in Europe, bears should probably stay in theoretical porridge.

The British reintroduction of the beaver pales in comparison to an ambitious and controversial project happening right now in the Netherlands. It is at the heart of discussions about our second R: rewilding. Whereas reintroduction is bringing back members of the same species recently lost due to definite human factors, rewilding is subtly different. Rewilding brings back the next best thing when wild species have been driven to complete extinction – so, cattle instead of extinct aurochs, feral horses instead of extinct wild horses. The theory behind rewilding as a force is that the ecological effects are what really matter, and not which particular species are producing them. Rewilding shifts the baseline not to any mythical golden age within human history but to a period before our species was on the scene.

This European rewilding has started in Oostvaardersplassen, a nature reserve close to Amsterdam. Oostvaardersplassen was reclaimed from the North Sea by the Netherlands in 1968. A polder set aside for new industrial development, it has instead become something much different – a giant ecological experiment in 6,000 hectares (15,000 acres) of former seabed. The same underwater space that was littered with the bones of megafauna from the Ice Age has now become home to modern analogues of aurochs and their kin. Spearheaded by Dutch ecologist Frans Vera, it has been stocked with Heck cattle, Konik ponies and red deer. Large grazers have kept the land open and steppe-like with few trees. They've recycled nutrients and manured the land. This has acted as a bold provocation to the idea that Europe was one big forest in the past, showing that the effects of large herbivores would have been significant in keeping some parts more like open veld.

Oostvaardersplassen's wilderness has even encouraged some of Europe's rarer inhabitants to move in of their own accord. White-tailed eagles nested in Oost-vaardersplassen a few years after the project was set up. They were the first Dutch white-tails to breed since the eighteenth century. A cinereous vulture, a giant scavenger more at home in the Dinaric Alps or Sierra Nevada, also patronised the Oostvaardersplassen. Cinereous vultures haven't been known since Roman times in the Netherlands. The most recent visitor was run over by a train when it strayed out of the reserve and into the twenty-first century. Because the Oostvaardersplassen is mostly left to run on its own schedule, there are always plenty of large mammal carcasses for raptors and scavengers to find.

It is a finite space with no large predators, hence starvation and death are the net result. Up to half of all the cattle, ponies and deer die each year, at least in part because feeble individuals that should become prey hoover up forage that could help sustain stronger survivors. More if the winter is harsh. Readily available carrion also means that other native species have moved back in. Red foxes and ravens pick bones clean and gorge on bloated corpses. The boom and bust cycles of growth and starvation are a public relations nightmare for the preserve. Visitors and supporters do not want to see emaciated animals on the point of death, however natural that death may be. It seems that people can put up with nature but not too much. Now, trained professionals are licensed to cull animals deemed unlikely to survive. Their bodies are left where they fall, feeding back into the ecosystem. I don't really know what is actually accomplished by the intervention. It's good for public relations, but death by bullet or death by starvation is not much of a choice. Again, the lack of lynx, wolf and bear becomes obvious and even cruel. Perhaps soon, wolves will naturally filter into the Oostvaardersplassen from some of their strongholds, as they tried to do in Denmark. There is still hope that the Dutch authorities might authorise the release of lynxes into the preserve too. Then we will see what top-down effects a complete trophic cycle will have; a new landscape of fear in this Serengeti behind the dykes.

Oostvaardersplassen only goes so far. There are no bison or moose (yet), and of course no woolly mammoths or woolly rhinos. As for the cattle and ponies, these are already familiar to most Dutch folk, seeing as they live happily amongst the windmills on farms and smallholdings throughout the country. One of the criticisms levelled at the whole project is that it's not much more than an abandoned farm. A different project, a continent away, has a much more ambitious remit: nothing less than bringing back the mammoth steppe and saving us from the perilous effects of human-induced global warming.

Pleistocene Park is the home and laboratory of Sergey Zimov. Since the 1980s he has been in charge of 15 square kilometres (6 square miles) of Yakutian tundra in the far north of the Kolyma river system, near the city of Chersky. Since then, Zimov has brought in hardy Yakutian horses, musk oxen, wisents and yaks to complement the concentrations of moose and reindeer already living in the region. For the first time since the early Holocene, Zimov has brought a suite of grazers back to the far north to observe what effects these large mammals have on the tundra ecosystem. The results have been spectacular. Where before there had been stunted larch, willow and mosses, the typical flora of the tundra, the work of the grazers has swept them away and allowed vibrant grasses to grow – the mammoth steppe grassland reappearing before their very eyes. Like the mammoths before them, the large grazers have made their own habitat.

Huzzah! Zimov's work shows quite clearly that mammals make the steppe, not the climate. And this has important repercussions for us, not just academically. The permafrost of Yakutia and Siberia is one of the biggest carbon sinks on the planet. Within the frozen soil lies preserved the organic gunk from tens of thousands of years. For now it is safely sequestered away. If the world warms much more, that organic mulch will be exposed to the surface and will provide fodder for carbon dioxide-producing microbes. Additionally, while it has been capped underground, anaerobic microbes have already been busy breaking it down into methane. As the permafrost thaws, that methane will be released, and methane is an infinitely more potent greenhouse gas than carbon dioxide. These are not trivial issues. Calculations by Sergey Zimov and colleagues suggest that up to 900 billion tonnes of carbon are sitting locked in the permafrost. That's more than the amount of carbon that has been released by humans burning fossil fuels since the start of the Industrial Revolution. A few degrees of warming, and the potential for a runaway warming feedback cycle is immense.

Zimov's Pleistocene Park could be one way to prevent this. Bringing back the mammoth steppe would help in two different ways. Having a covering of grass instead of moss and shrubs would reflect more sunshine in the long Arctic summer and help to insulate the frozen ground underneath. In winter, when everything is covered in snow, mashing down the snow cover under the hooves of megafauna compacts it, making it a worse insulator and keeping the ground frozen. Counter-intuitively, snow usually helps to warm the ground. Like a cosy igloo, the air within fluffy snow on the ground insulates and allows temperatures to rise, melting the permafrost underneath. When grazing wisents and horses clear the snow, the freezing Arctic air is allowed to circulate and keep temperatures lower.

Pleistocene Park is a broader experiment than Oostvaardersplassen in another important dimension. Carnivores are present. Northern lynxes, wolves and brown bears are always sniffing around the edges and have taken their fair share of animals. It helps to be so far away from a sensitive and squeamish public, but Pleistocene Park does not have any public relation issues with regards to starving animals. The wolves get them. Sergey Zimov has managed to bring back a fair approximation of a working Pleistocene ecosystem but in the Holocene.

The greatest success of Zimov's work is showing that the ecological services that megafauna provided are not trivial, and that these ecological services are sorely lacking in many of the defaunated regions of the world. Not just to protect soils or increase diversity but to restore what once was there. Does it not make sense, then, to try to restore some of those ecological interactions using close counterparts where species have gone extinct? This is exactly what some ecologists have been arguing for. They've even started to do it.

Mauritius, the island in the middle of the Indian Ocean that was famously home to the extinct dodo, didn't just lose this enormous flightless pigeon. A whole ecosystem of birds found their way to extinction via humans, pigs and dogs. The island was also home to a couple of species of giant tortoises that were similarly exploited into oblivion. Now, Mauritius has very few native species left and is brimful of introduced species, not least of which is sugarcane. The native flora has suffered too, bereft of the birds and animals that evolved to disperse their seeds and eat their fruit. Mauritius used to have a plethora of species of ebony hardwoods, but time and logging have deforested most of the island and few mature trees have grown back.

One species of Mauritian ebony (*Diospyros egrettarum*) is critically endangered. In a bold plan, researchers introduced an exotic species of Aldabran giant tortoise onto the small Île aux Aigrettes to monitor the effects that rewilding with tortoises could have. Bear in mind that this particular species had never lived on Mauritius, belonged to a completely different genus from the native Mascarene tortoise, and had not evolved to eat ebony seeds of this kind. It was the ecological equivalent of bringing an elephant into Pleistocene Park and expecting it to do the job of the mammoths of old. Amazingly, the Mauritian tortoise rewilding was an enormous success. The transplanted Aldabran giant tortoises fulfilled the niche that their extinct relatives had left unfilled for centuries. They ate the ebony seeds and dispersed them away from their parent trees, giving them a chance to germinate out of the shade, in new spots, and with a lovely dollop of fertilising tortoise droppings to see them on their way. What's more, passage through the tortoise massively improved the seeds' potential for germination. They sprouted earlier and were more likely to sprout at all compared to seeds that hadn't been ingested. It's probable that the ebony seeds had evolved in tandem with the extinct Mauritian tortoises, and that passage through the gut signalled to the seeds that they were in a good position to grow. With the replacement Aldabran tortoises, that particular ecological interaction had been restored and the critically endangered ebony could form seedlings again naturally. Fewer than a dozen *Diospyros egrettarum* ebonies still grow on the main island of Mauritius. Tortoises could be their saviour.

The idea that many plants have evolved in a symbiotic relationship with their dispersers is not a new one. What is relatively new is the recognition that many of the dispersing partners are probably extinct. Avocado and cocoa trees are just two of the many American flora that we think were dispersed by ground sloths, mastodons or some other Pleistocene megafauna. Without human help, they would both likely have gone extinct. Where they grow naturally in South America, the largest surviving mammal is the tapir, which is certainly not big enough to handle an avocado seed or cocoa pod.

What all this points to is that rewilding is an exciting, credible and innovative way to help nature bring back some of the balance that has been lost. It's been called 'punk ecology', and that's how I like to think of it.

There is a flipside, though, showing that rewilding has some dangers. What are invasive species if not examples of rewilding gone wrong? The crazy transplantations of European bird species to Australia, New Zealand and the Americas by blasted acclimatisation societies have unleashed disasters like the house sparrow, which flat out kills native birds. Escapes from fur farms allowed the American mink to colonise Britain and drive the native water vole to the brink of extinction. Feral dogs and cats[*] all over the world are responsible for almost as

---

[*] Hands down the saddest extinction of recent times is of the playful little flightless Lyall's wren. Once widespread over New Zealand, by the time of European colonisation it only lived on Stephens Island in the Cook Strait. This small rat-free island was its refuge until a lighthouse was built and people brought cats as companions. The cats escaped and one, named Tibbles, brought the defenceless wrens as gifts to the lighthouse keeper. From 1894 to 1895, over a dozen small bodies were deposited on the lighthouse doorstep and passed by the keepers on to museum collectors. Then Tibbles stopped bringing them in. So died the only flightless passerine of modern times. Some say that the Twitter logo pays homage to this extinct, flightless little mouse of a songbird.

many modern extinctions as we are. Basically, moving species around is an extremely hazardous business and not something we should rush into. Certainly not with anything that has the slightest carnivorous tendencies.

What we can do is learn from accidental rewilding and invasive species we have moved around the globe and see if there are any lessons to be learned there. Australia is a good focus. Of all the continents, Australia has been stripped of megafauna for the longest period of time. Modern humans got there early and the native giant marsupials disappeared. That isn't to say that Australia is lacking in megafauna now. It's just not native. At all. I'm talking feral dromedary* camels, cows, brumbies, banteng, water buffaloes, donkeys, sambar deer, red deer, fallow deer, rusa deer and feral boars. For species of megafauna, Australia is as rich now as it was in the Pleistocene; it's just that none of them evolved in Australia.

Is that a bad thing? Yes and no. Dromedaries are extinct in the wild. Banteng (wild Asian bovids) are endangered. The Aussie populations act as ecological surrogates and back-up populations free of human interference. Over time they have started a process of integrating into their new home. As proof of this, an amazing and novel relationship has grown up between the new and the old. Native Australian crows now offer grooming services to banteng, a species that's only been in their habitat for a century. Like the buffalo with the

---

* One-humpers. I find it easy to remember which camel is which by looking at the starting letter. Dromedary (D) has one hump. Bactrian (B) has two humps!

oxpecker, banteng have started to interact with the crows and even lie on their back with legs held high to allow the crows access to their most annoying ticks. The crows could be reverting to behaviours last practised on giant Pleistocene wombats or they could have worked out how to get an easy meal from scratch. It might be the start of a beautiful symbiosis.

As for the other species' presence in Australia, that's a mixed bag. Brumbies and feral boars are culled due to their damaging of native flora. The deer are hunted for sport. It's very tricky because none of the species were purposefully released; they are all escapees who made a break for freedom from their roles as pack or farm animals. The Australian government is keeping a very close eye on all of them and is monitoring what effects they have on the delicate balance of native ecosystems. Some may end up being as beneficial as Mauritian tortoises, others are definitely not welcome.* Time will tell.

Of course, discussion of rewilding becomes a moot point if we can achieve our third R: resurrection. No need for Aldabran tortoises on Mauritius if you can somehow bring back the extinct Mauritius species and use them. Generally, I prefer the slightly romantic term of 'resurrection' to the more commonly used

---

* This is true in other countries as well. Colombia has a problem with feral hippos. The well-known narco Pablo Escobar ran a small zoo from his home, Hacienda Nápoles, to fuel his Robin Hood persona. After he was shot dead by police, the hippos escaped and made their home in the region. There are now around 30 of the fearsome beasts and they are known to attack fisherfolk and cattle.

'de-extinction'. Resurrection has the subtle implication of arcane magic and suggests that whatever is brought back will have been changed by the process. De-extinction has been popularised by my mentor Beth Shapiro, and adopted as an unwieldy but descriptive heading by many. It's a word that's needed, because every time a new scientific paper on the DNA of extinct species is published, or a new permafrost mummy is unearthed, the press have a field day with stories of how scientists are working on cloning a woolly mammoth, a cave lion, a woolly rhino or whatever. However interesting the work is on ancient DNA, only one lead is ever used; only one question is ever asked. When are we going to clone them?

Most people aren't really aware that scientists have cloned an extinct species already. It's not generally recognised, perhaps because the resurrection happened with a woolly goat rather than a woolly mammoth. It was a phenomenal breakthrough, and deserves to be better known. The bucardo, or Pyrenean ibex, was a majestic (for a goat) mountain animal, with those sweeping ridged horns that you only ever see mounted over some toff's fireplace. The population from the north of Spain was intensely hunted until it entered an extinction vortex that scientists could only watch and monitor from the sidelines. Thus, it came to pass that the very last individual, a female named Celia, was found dead under a fallen tree on 6 January 2000.

Extinction wasn't forever, though. While she was still alive, Celia had tissue samples taken and stored in deep-freeze as an insurance policy. For three years they sat in a bath of liquid nitrogen, the only tissue from a species newly extinct. Using the same techniques pioneered on cloning Dolly the sheep, Spanish scientists carefully took the nuclei of Celia's cells and transplanted them into empty egg cells from domestic goats. Then, 154 bucardo embryos were implanted into 44 nanny goats, resulting in just five pregnancies. Only one proceeded to term. On Wednesday, 30 July 2003, after an extinction of three years, a baby bucardo was resurrected. But only for a few minutes. Tragically, the newborn clone died soon after birth from a lung abnormality. Because of this, the bucardo has the distinction of being the only species to have gone extinct twice. Researchers are still attempting to bring viable bucardo clones back to life.

The laboratory technique used on Dolly and Celia, known as somatic cell nuclear transfer, is notoriously fiddly. It is, however, currently the only protocol for cloning mammals. It's never going to be workable with Pleistocene megafauna, even those preserved in the permafrost, because it requires living cells to work. However, much of the excitement surrounding studies of ancient DNA focuses on how many complete nuclear genomes have been produced. And surely once you have the genome it should be child's play to bring the species back? Unfortunately, that's not the case. When we talk about complete genomes, from mammoths, from cave bears, from Neanderthals, from aurochs, they only exist *in silico*. Ancient genomes require huge amounts of computing power to reassemble what is basically millions of broken down DNA fragments into a coherent whole. Like a shredded book they have to be

reassembled to make sense. Except it's more difficult
even than that. Ancient genomes need a modern scaffold
to act as guide to reassembly. In each case, a close living
relative is needed to map the layout of the genes.
Modern elephant genomes are used to reconstruct
mammoths; modern humans for Neanderthals. Imagine
trying to reconstruct the pages of *Hamlet* with only
Rosencrantz and Guildenstern as a guide, and that's
basically what happens when ancient genomes are
mapped. Because it's done in this way, vast swathes of
sequence information are missing, including critical
aspects of the architecture of the chromosome. Like
mammoth somatic cell nuclear transfer, the idea of
sequencing a mammoth genome and popping it into a
cell is also a pipe dream.

There is, however, one avenue that's recently opened
up in the quest to resurrect extinct species. And it just
might work to revive some of the Pleistocene megafauna.
Instead of trying to find living cells, or synthesise
recreations of the genomes of extinct species, why not
tweak the living cells of their surviving relatives to
produce a facsimile of what's missing? Could we use the
hard-won genomes of extinct woolly mammoths to
identify where they diverge from Asian elephants and
just swap the sections over?

We are already at that stage. Right now, at a small lab
at Harvard Medical School, sitting in a petri dish are
elephant cells with mammoth genes spliced into them.
The lab of George Church has proceeded into the realms
of science fiction by taking an alternative approach to
species resurrection. Reasoning that what makes a
mammoth special is its physical characteristics, the

Church lab has identified differences in the genomes between elephants and mammoths that control things like hair length, ear size and fat metabolism. Alongside the gene for the mammoth's unique haemoglobin and a handful of other candidate genes, Church's group have used molecular biology to engineer elephant cells that are a little bit mammothy. For the first time since Wrangel Island, mammoth genes are growing and replicating in living cells in an incubator near Boston, Massachusetts. How incredible is that?

The production of elephant-mammoth cells used a technology called CRISPR/cas9. CRISPR/cas9 is a relatively new molecular biology technique stolen from the primitive immune responses of bacterial cells. When bacteria are infected with viruses, the CRISPR/cas9 complex of proteins interacts with the foreign viral genes and acts like a heat-seeking missile to find other similar snatches of the virus gene and cut them up. In infected bacteria, this stops the viruses from reproducing and gives the bacteria a chance to live another day. The beauty of the system is that it is programmable – any stretch of gene can be used to tell the CRISPR/cas9 what to seek out and cut. Of course, this acts as a perfect tool for molecular biologists, who have long had the ability to produce DNA sequences on demand for use in experiments. The brilliance of the technique is that it can be used on living cells to target specific gene sections and cut out exact pieces of DNA to be swapped with other pieces. It's the perfect nanobot for all kinds of gene therapy. For the first time, scientists have the right tool to target individual genes *in vivo* and change them however they want. And what they want is to make elephant cells that express mammoth genes!

Those precious cells in the Church lab are an amazing first step in the resurrection of the mammoth. It's a Lyuba-sized step, though. The Harvard lab has made about 14 changes to its Asian elephant cells. Asian elephants and woolly mammoths differ at around 400,000 places in their respective genomes. It all comes down to how much mammoth you want in your elephant. The 400,000 differences between elephants and mammoths will not all code for physical differences. Many will be natural, neutral examples of the mammoths' genetic diversity. Still, the Church lab has a long way to go until it produces something that even approaches a totally mammoth cell.

When that happens, it might be possible to then use somatic cell nuclear transfer to put the nucleus into an elephant egg (maybe generated from stem cells). But even at this advanced stage there would be intractable problems. Elephant IVF is a dangerous business. Elephants only ovulate twice a decade and in the wild almost immediately become pregnant with foetuses that they carry for nearly two years. No elephant babies have been born with IVF because elephant hymens are impervious to anything larger than sperm cells. Even after calves are born, the hymen grows back as an extra barrier to human tinkering.

It's an understatement indeed to say that the road to resurrecting mammoths, or woolly rhinos, or cave bears, or sabretooths, is a long and winding one. Apart from all the technical problems, we can't afford to neglect the ethical issues that having such god-like powers provokes. Supposing one day, years from now, we do resurrect the mammoth. What would we do with it? It would be alone. It would be a mammoth in a world that now belongs to elephants. It would have no culture to learn. It would live its life in captivity, an ugly duckling never able to live as a swan. Supposing we could create multiple mammoths, male and female. What would be the end plan? To release them in Pleistocene Park and help turn the tundra back into steppe? I'm sceptical. Bringing mammoths back just to try and fix problems we've caused makes it sound like we've learned nothing from their extinction in the first place. Engineering mammoth-like elephants and semi-woolly rhinos to work for us, offsetting our carbon debt sounds horrific. To bring such majestic animals back to be gawked at in a zoo or to bioengineer grassland seems like a mistake to me. Nobody would be more thrilled to see a living sabretooth or cave lion than I would but not at that price. I'd rather live with their memory than see them corrupted into living exhibits or beasts of burden.

There is a fourth R that I didn't mention, and which may be the most important of them all: remembrance. If I've done anything with this book, it's to emphasise that extinction is not something safely confined to the past. There is no cut-off point, no box where the spectre of human-caused extinction can be confined. As long as our species has been on the planet, we have caused extinction and we still continue to cause it. After all, this world can be a fragile place, and our group actions can have massive consequences. For nearly 20 years I have learned everything I could about the extinct species of the Pleistocene and Holocene to better understand where we are now. Remembrance can be a powerful force when it is used to promise 'never again'. I would like to think that by understanding why some species have gone extinct, we could all do better going forward to avoid repeating the mistakes of the past.

Afterword

I'd like to end the book on a personal note. Writing this story of extinction has been a cathartic experience for me. All my professional life I've been somewhat obsessed with the life of the past. I've spent thousands of hours hunched over test tubes, in aseptic clean rooms, and looking at strings of DNA on computer screens. Until writing this book I had no real idea what drove me to spend so much of my waking life trying to amplify DNA from extinct beasts. Over the past year I had something of an epiphany. The planning and drafting stages allowed for plenty of navel-gazing and introspection, and it all of a sudden dawned on me why I might be the way I am. If you are a reader who has persevered this far, I think I owe you some kind of explanation.

When I was a seven-year-old boy, my mother died. It was unexpected, to me, because even though I knew she was ill, the severity of that illness hadn't reached me. My parents had done their best to shield me and my sister from their own pain. As my new reality warped around me, I remember thinking that I never wanted to lack that kind of understanding ever again. As time shifted forward I readjusted to the prospect of a different kind of life. It's only now, 30 years later and with some armchair psychology, that I think I can see a connection between the boy I was and the man I am. All those years spent learning how to resurrect DNA from the Pleistocene casts me as aspiring to be a 'modern Prometheus'. I see now that my obsession probably stems from a desire to bring the past back to life.

Loss is a very difficult concept to put into words for those who have not experienced it first-hand. At first it's like your whole self has been hollowed out. Later, you find things to fill that space. For me, biology and the natural world replenished my sense of self.

Now, I never feel so connected to nature as when I ponder what we've lost.

Appendix

This appendix provides a list of all the scientific names of species referred to in this book.

But first, a quick explanation of how scientists deal with species and, in particular, their names. To know which species you are referring to, the binomial (Latin name) system is unambiguous if a tad unwieldy. It can be split into two parts: the first is the genus (plural genera) and denotes that all members of that group have a recent common ancestor. The genus always gets a capital letter. The genus *Panthera* is a good example: lions, leopards, jaguars, tigers and snow leopards are all obvious close relatives that share large size, the ability to roar, patterned coats and many details of the skeleton. They all fit into one genus and have a direct ancestor that links them all. The second part is the species name (e.g. *leo, pardus, onca, tigris* or *uncia*). It always starts with a lower case. Species themselves are not static blocks. They can be further subdivided into subspecies that represent different pockets of diversity and adaptations to different habitats. A good example of this is in tigers (*Panthera tigris*). The enormous, long-haired, cold-adapted tiger that lives in Pacific Russia, the Amur or Siberian tiger, is *Panthera tigris altaica*. The small, sleek form that lives on the island of Sumatra is *Panthera tigris sumatrae*. With the addition of a subspecies name, you get information not just on the species and genus but also the geographic region where the animal comes from.

At the other end of the scale, different genera will belong to the same taxonomic family and share a common ancestor even further back in time. The big cats (*Panthera* spp.), the small cats (*Felis* spp.) and the cheetah (*Acinonyx* sp.) belong to the Felidae, a family that can easily be understood as containing all the felids, and descended from a cat that lived in the Late Miocene (about 11 million years ago). In this instance, the taxonomic shorthand 'spp.' refers to multiple species within a genus, whereas 'sp.' means there is only one referred species in the genus.

2-toed sloth	*Choloepus* spp.
3-toed sloth	*Bradypus* spp.
Aardwolf	*Proteles cristatus*
African cheetah	*Acinonyx jubatus*
African elephants	*Loxodonta* spp.
Alaskan brown bear	*Ursus arctos middendorffi*
Aldabran giant tortoise	*Aldabrachelys gigantea*
American beaver	*Castor canadensis*
American buffalo	*Bison bison*
American cheetah	*Miracinonyx trumani*
American lion	*Panthera atrox*
American mastodon	*Mammut americanum*
American mink	*Neovison vison*
Ancestral cave lion	*Panthera leo fossilis*
Archaeoindris	*Archaeoindris fontoynontii*
Archaeolemur	*Archaeolemur* spp.
Arctic fox	*Vulpes lagopus*
Ash	*Fraxinus* spp.
Asian elephant	*Elephas maximus*
Aspen	*Populus tremuloides*

Atlantic salmon	*Salmo salar*
Atlas bear	*Ursus arctos crowtheri*
Aurochs	*Bos primigenius*
Avocado	*Persea americana*
Bactrian camel	*Camelus bactrianus*
Banteng	*Bos javanicus*
Barbary lion	*Panthera leo leo*
Beaver	*Castor* spp.
Beringian cave lion	*Panthera spelaea*
Bison	*Bison* spp.
Black bear	*Ursus americanus*
Black-backed jackal	*Canis mesomelas*
Blue fox	*Vulpes lagopus*
Bluebottle	*Calliphora vomitoria*
Boar	*Sus scrofa*
Bobcat	*Lynx rufus*
Brown bear	*Ursus arctos*
Brown hyaena	*Parahyaena brunnea*
Brumbies	*Equus ferus caballus*
Bucardo	*Capra pyrenaica pyrenaica*
Bush dog	*Speothos venaticus*
Camels	*Camelus* spp.
Canada lynx	*Lynx canadensis*
Cane toad	*Bufo marinus*
Cape buffalo	*Syncerus caffer*
Cape lion	*Panthera leo melanochaita*
Capybara	*Hydrochoerus hydrochaeris*
Caracal	*Caracal caracal*
Caribou	*Rangifer tarandus*
Carpet beetles	*Anthrenus* spp.
Caspian tiger	*Panthera tigris virgata*

Cassowaries	*Casuarius* spp.
Castoroides	*Castoroides ohioensis*
Cattle	*Bos primigenius taurus*
Cave hyaena	*Crocuta crocuta spelaea*
Cave lion	*Panthera spelaea*
Chamois	*Rupicapra rupicapra*
Cheetah	*Acinonyx* spp.
Chicken	*Gallus gallus domesticus*
Chimpanzee	*Pan troglodytes*
Chytridiomycosis	*Batrachochytrium dendrobatidis*
Cinereous vulture	*Aegypius monachus*
Clouded leopard	*Neofelis nebulosa*
Cobras	*Naja* spp.
Cocoa	*Theobroma cacao*
Cod	*Gadus* spp.
Columbian mammoth	*Mammuthus columbi*
Conturines cave bear	*Ursus ladinicus*
Cottonwood	*Populus* spp.
Cougar	*Puma concolor*
Cow	*Bos primigenius taurus*
Coyote	*Canis latrans*
Crab-eating fox	*Cerdocyon thous*
Cuvier's gomphothere	*Cuvieronius hyodon*
Darwin's fox	*Lycalopex fulvipes*
Darwin's ground sloth	*Mylodon darwinii*
Deep-nosed horse	*Hippidion saldiasi*
Deninger's cave bear	*Ursus deningeri*
Denisovan	*Homo x*
Dire wolf	*Canis dirus*
Dirk-toothed cat	*Smilodon populator*

Dodo	*Raphus cucullatus*
Dog	*Canis lupus familiaris*
Domestic cat	*Felis silvestris catus*
Domestic goat	*Capra aegagrus hircus*
Donkey	*Equus africanus, Equus africanus asinus*
Dormouse	*Muscardinus avellanarius*
Dromedary	*Camelus dromedarius*
Dugong	*Dugong dugon*
Emu	*Dromaius novaehollandiae*
Eurasian beaver	*Castor fiber*
Eurasian giant cheetah	*Acinonyx pardinensis*
European aurochs	*Bos primigenius primigenius*
European badger	*Meles meles*
European bison	*Bison bonasus*
European elk	*Alces alces*
European jaguar	*Panthera gombaszoegensis*
Falkland Islands wolf	*Dusicyon australis*
Fallow deer	*Dama dama*
Feral boar	*Sus scrofa*
Feral horse	*Equus ferus caballus*
Field vole	*Microtus agrestis*
Florida panther	*Puma concolor coryi*
Forest elephant	*Loxodonta cyclotis*
Gamssulzen cave bear	*Ursus ingressus*
Gaur	*Bos gaurus*
Giant American beaver	*Castoroides ohioensis*
Giant anteater	*Myrmecophaga tridactyla*
Giant deer	*Megaloceros giganteus*
Giant elephant bird	*Aepyornis maximus*

Giant koala lemur	*Megaladapis edwardsi*
Giant llama	*Llama gracilis*
Giant Mascarene tortoises	*Cylindraspis inepta,*
	Cylindraspis spp.,
	Cylindraspis triserrata
Giant moa	*Dinornis* spp.
Giant New Zealand eagle	*Harpagornis moorei*
Giant panda	*Ailuropoda melanoleuca*
Giant Patagonian jaguar	*Panthera onca mesembrina*
Giant ripper	*Megalania prisca*
Giant short-faced kangaroo	*Procoptodon goliah*
Giant sloth lemur	*Archaeoindris fontoynontii*
Giant sloths	*Megatherium americanum,*
	Mylodon darwinii,
	Nothrotheriops shastensis,
	Paramylodon harlani
Giant wombat	*Diprotodon optatum*
Gir lion	*Panthera leo persica*
Giraffe seahorse	*Hippocampus*
	camelopardalis
Giraffes	*Giraffa* spp.
Glyptodonts	*Doedicurus clavicaudatus,*
	Glyptodon clavipes,
	Panochthus spp.
Golden eagle	*Aquila chrysaetos*
Golden jackal	*Canis aureus*
Gomphotheres	*Cuvieronius hyodon,*
	Notiomastodon platensis
Gorilla	*Gorilla gorilla*
Great auk	*Pinguinus impennis*
Grey squirrel	*Sciurus carolinensis*
Grey wolf	*Canis lupus*

Grizzly bear	*Ursus arctos*
Ground sloths	*Megatherium americanum, Mylodon darwinii, Nothrotheriops shastensis, Paramylodon harlani*
Haddock	*Melanogrammus aeglefinus*
Harington's horse	*Haringtonhippus francisci*
Hazel	*Corylus* spp.
Hippopotamus	*Hippopotamus amphibius*
Horned tortoises	*Meiolania platyceps, Ninjemys oweni*
Horses	*Equus ferus*
House sparrow	*Passer domesticus*
Human	*Homo neanderthalensis, Homo sapiens*
Humped cattle	*Bos primigenius indicus*
Iberian lynx	*Lynx pardinus*
Ibex	*Capra ibex*
Indicine cattle	*Bos primigenius indicus*
Irish elk	*Megaloceros giganteus*
Issoire lynx	*Lynx issiodorensis*
Jackals	*Canis adustus, Canis aureus, Canis mesomelas*
Jaguar	*Panthera onca*
Kangaroos	*Macropus* spp.
Kiwis	*Apteryx* spp.
Koala	*Phascolarctos cinereus*
Larch	*Larix* sp.
Large Cuban sloth	*Megalocnus rodens*
Leopard	*Panthera pardus*
Leopard cat	*Prionailurus bengalensis*
Lesser elephant bird	*Mullerornis agilis*

Lion	*Panthera leo*
Litoptern	*Macrauchenia patachonica*
Llama	*Lama glama*
Lord Howe Island horned tortoise	*Meiolania platyceps*
Lyall's wren	*Traversia lyalli*
Madagascar pygmy hippo	*Hippopotamus lemerlei*
Mammoths	*Mammuthus* spp.
Manatee	*Trichechus manatus*
Maned wolf	*Chrysocyon brachyurus*
Marsupial lion	*Thylacoleo carnifex*
Mastodon	*Mammut americanum*
Mauritian broad-billed parrot	*Lophopsittacus mauritianus*
Mauritian ebony	*Diospyros egrettarum*
Mauritian tortoise	*Cylindraspis triserrata*
Medium-sized Caribbean sloth	*Neocnus comes*
Megaladapis	*Megaladapis edwardsi*
Mexican horse	*Equus conversidens*
Middle Eastern lion	*Panthera leo persica*
Moa	Order Dinornithiformes
Modern lion	*Panthera leo*
Moon bear	*Ursus thibetanus*
Moose	*Alces alces*
Morning star glyptodont	*Doedicurus clavicaudatus*
Mosbach horse	*Equus mosbachensis*
Mosbach wolf	*Canis mosbachensis*
Mountain lion	*Puma concolor*
Musk ox	*Ovibos moschatus*
Neanderthal	*Homo neanderthalensis*
North American elk	*Cervus canadensis*
Northern lynx	*Lynx lynx*
Northern tamandua	*Tamandua mexicana*

Ocelot	*Leopardus pardalis*
Ostrich	*Struthio camelus*
Otter	*Lutra lutra*
Owen's ninja turtle	*Ninjemys oweni*
Oxpecker	*Buphagus* spp.
Palaeopropithecus	*Palaeopropithecus ingens*
Pallas's cat	*Otocolobus manul*
Pallas's cormorant	*Phalacrocorax perspicillatus*
Pampatheres	*Pampatherium* spp.
Passenger pigeon	*Ectopistes migratorius*
Patagonian short-faced bear	*Arctotherium tarijense*
Pigs	*Sus* spp.
Pine marten	*Martes martes*
Polar bear	*Ursus maritimus*
Poplar	*Populus* spp.
Possums	*Didelphis* spp.
Puffin	*Fratercula arctica*
Puma	*Puma concolor*
Pyrenean ibex	*Capra pyrenaica pyrenaica*
Rabbit	*Oryctolagus cuniculus*
Raccoon dog	*Nyctereutes procyonoides*
Rainbow snake	*Wonambi naracoortensis*
Ramesch cave bear	*Ursus eremus*
Raven	*Corvus corax*
Razorbill	*Alca torda*
Red deer	*Cervus elaphus*
Red fox	*Vulpes vulpes*
Red squirrel	*Sciurus vulgaris*
Red wolf	*Canis rufus*
Reindeer	*Rangifer tarandus*
River otter	*Lutra lutra*

Roe deer   *Capreolus capreolus*
Rusa deer   *Rusa timorensis*
Sabretooth cat   Family Felidae,
                                Subfamily
                                Machairodontinae

Sambar deer   *Rusa unicolor*
Savanna elephant   *Loxodonta africana*
Scimitar cat   *Homotherium latidens*
Scimitar-toothed cat   *Homotherium latidens*
Scottish wildcat   *Felis silvestris grampia*
Sea otter   *Enhydra lutris*
Serval   *Leptailurus serval*
Shasta ground sloth   *Nothrotheriops shastensis*
Sheep   *Ovis aries*
Shelk   *Megaloceros giganteus*
Short-faced bear   *Arctodus simus*
Shoveler   *Anas clypeata*
Siberian tiger   *Panthera tigris altaica*
Side-striped jackal   *Canis adustus*
Sloth bear   *Melursus ursinus*
Sloth lemur   *Palaeopropithecus* spp.
Sloths   Order Pilosa, Suborder
                                Folivora

Small cave bear   *Ursus rossicus*
Snow leopard   *Panthera uncia*
South Island giant moa   *Dinornis robustus*
Southern beech   *Nothofagus* spp.
Southern gomphothere   *Notiomastodon platensis*
Spectacled bear   *Tremarctos ornatus*
Spotted hyaena   *Crocuta crocuta*
Stag-moose   *Cervalces scotti*

Stegodonts	*Stegodon* spp.
Steller's eider	*Polysticta stelleri*
Steller's jay	*Cyanocitta stelleri*
Steller's sea cow	*Hydrodamalis gigas*
Steller's sea eagle	*Haliaeetus pelagicus*
Steller's sea lion	*Eumetopias jubatus*
Steppe bison	*Bison priscus*
Steppe mammoth	*Mammuthus trogontherii*
Striped hyaena	*Hyaena hyaena*
Sugarcane	*Saccharum* sp.
Sumatran rhinoceros	*Dicerorhinus sumatrensis*
Sumatran tiger	*Panthera tigris sumatrae*
Sun bear	*Helarctos malayanus*
Sundaland clouded leopard	*Neofelis diardi*
Tapir	*Tapirus* spp.
Tasmanian tiger	*Thylacinus cynocephalus*
Thylacine	*Thylacinus cynocephalus*
Tiger	*Panthera tigris*
Torresian crow	*Corvus orru*
Toxodonts	*Toxodon* spp.
Tretretretre	*Palaeopropithecus* spp.
Trogontherium	*Trogontherium cuvieri*
Trout	*Salmo trutta*
Walrus	*Odobenus rosmarus*
Wapiti	*Cervus canadensis*
Warthogs	*Phacochoerus* spp.
Water buffalo	*Bubalus bubalis*
Water vole	*Arvicola amphibius*
Weasel	*Mustela erminea*
White rhinoceros	*Ceratotherium simum*
White-nose syndrome	*Geomyces destructans*

White-tailed eagle	*Haliaeetus albicilla*
Wild boar	*Sus scrofa*
Wild horse	*Equus ferus ferus*
Wildebeest	*Connochaetes taurinus*
Willow	*Salix* spp.
Wisent	*Bison bonasus*
Wolf	*Canis lupus*
Wolverine	*Gulo gulo*
Wombat	*Vombatus* sp., *Lasiorhinus* spp.
Woolly mammoth	*Mammuthus primigenius*
Woolly rhinoceros	*Coelodonta antiquitatis*
Yak	*Bos grunniens*
Yakutian horse	*Equus ferus caballus*
Yesterday's camel	*Camelops hesternus*
Zebras	*Equus* spp.
Zebu	*Bos taurus indicus*

Acknowledgements

Thanks primarily to Jim Martin and Jenny Campbell at Bloomsbury for everything they have done.

Thanks also to Myriam Birch for her word-wrangling.

Many people have helped me on my journey, and a proper list of those I am indebted to would be longer than the book itself. For help with the research specifically for this book, I spoke to many experts in many fields: Ian Barnes, Jim Crumley, Love Dalén, Ceiridwen Edwards, David Hetherington, Andrew Kitchener, Greger Larson, Adrian Lister, Joyce Lundberg, Pavel Nikolskiy, Hannah O'Regan, Alison Parfitt, Patrícia Pečnerová, Malgorzata Pilot, Jelle Reumer, Ricardo Rodrígues-Varela, Beth Shapiro, Tony Stuart, Naomi Sykes, Becky Wragg Sykes, Cris Valdiosera and Lars Werdelin gave me invaluable insight into many species.

All maps used in this book are modified from originals from d-maps (www.d-maps.com).

I would like to thank my writing and blogging buddies Jan Freedman and Rena Maguire for copious kind words and encouragements. I'd also like to thank Anna Barker and Pat Barker for advice and motivation. Many friends past and present helped me on my way: Sinclair Swan, Iain Innes, Blair Duncan, Christopher Gell, Simon Lloyd, Simon Sylvester, Jordan Dunn, Simon Baker, Billy Kinnear and a number of others whose constant criticisms of my many personal faults prepared me for the rigours of publication.

Thanks to my family: my sisters Jaqueline, Adriana, Lexi; my parents Dave and Raquel. To my late, beloved grandparents, Nan and George, who did so much to make me who I am. Thanks to my mother-in-law, Elizabeth Howe, for support and babysitting.

Thank you to Alice and Naomi for showing me new worlds every day.

Bittersweet gratitude for my mother's memory, for giving me my first insight into extinction.

'Thanks' is too small a word to encompass my appreciation of my wife, Bridie, who has been with me through thick and thin for two decades.

Further Reading

A full list of resources can be found at researchgate.net/profile/
Ross_Barnett

Chapter 1: The Past

Goodman, S. M., Jungers, W. L., Simeonovski, V. (2014) *Extinct Madagascar*. Chicago: University of Chicago Press.

Grayson, D. K. (2016) *Giant Sloths and Sabertooth Cats*. Salt Lake City, Utah: University of Utah Press.

Guthrie, R. D. (1990) *Frozen Fauna of the Mammoth Steppe: The Story of Blue Babe*. London: University of Chicago Press.

Kolbert, E. (2014) *The Sixth Extinction: An Unnatural History*. London: Bloomsbury.

Kurtén, B. (1968) *Pleistocene Mammals of Europe*. London: Weidenfeld and Nicolson.

Kurtén, B., Anderson, E. (1980) *Pleistocene Mammals of North America*. New York: Columbia University Press.

Martin, P. S. (2005) *Twilight of the Mammoths: Ice Age Extinction and the Rewilding of America*. Berkeley: University of California Press.

Martin, P. S., Klein, R. G. (1984) *Quaternary Extinctions: A Prehistoric Revolution*. Tucson, Arizona: University of Arizona Press.

Steller, G. W. (1751) *De Bestiis Marinis, or, The Beasts of the Sea*. Washington DC: Government Printing Office.

Woodward, J. (2014) *The Ice Age: A Very Short Introduction*. Oxford: Oxford University Press.

Chapter 2: Cave Hyaena

Bon, C., Berthonaud, V., Maksud, F., Labadie, K., Poulain, J., Artiguenave, F., Wincker, P., Aury, J. M., Elalouf, J. M. (2012) Coprolites as a source of information on the genome and diet of the cave hyena. *Proceedings of the Royal Society B: Biological Sciences*, 279, 2825–2830.

Buckland, W. (1822) Account of an assemblage of fossil teeth and bones of elephant, rhinoceros, hippopotamus, bear, tiger, and hyaena, and sixteen other animals; discovered in a cave at Kirkdale, Yorkshire, in the year 1821: With a comparative view of five similar caverns in various parts of England, and others on the continent. *Philosophical Transaction of the Royal Society of London*, 112, 171–236.

Frank, L. G., Glickman, S. E. (1994) Giving birth through a penile clitoris: Parturition and dystocia in the spotted hyaena (*Crocuta crocuta*). *Journal of Zoology (London)*, 234, 659–690.

Glickman, S. E. (1995) The spotted hyena from Aristotle to the lion king: Reputation is everything. *Social Research*, 62, 501–537.

Kruuk, H. (1972) *The Spotted Hyena: A Study of Predation and Social Behaviour*. Chicago and London: The University of Chicago Press.

Rohland, N., Pollack, J. L., Nagel, D., Beauval, C., Airyaux, J., Pääbo, S., Hofreiter, M. (2005) The population history of extant and extinct Hyenas. *Molecular Biology and Evolution*, 22, 2435–2443.

Stuart, A. J., Lister, A. M. (2014) New radiocarbon evidence on the extirpation of the spotted hyaena (*Crocuta crocuta* (Erxl.)) in northern Eurasia. *Quaternary Science Reviews*, 96, 108–116.

Chapter 3: Sabretooth Cat

Antón, M. (2013) *Sabertooth*. Bloomington: Indiana University Press.

Barnett, R. (2014) An inventory of British remains of *Homotherium* (Mammalia, Carnivora, Felidae), with special reference to the material from Kent's Cavern. *Geobios*, 47, 19–29.

Dawkins, W. B., Sanford, W. A., Reynolds, S. H. (1866) *A Monograph of the British Pleistocene Mammalia*. London: Palaeontographical Society.

MacEnery, J. (1859) *Cavern Researches: Or, Discoveries of Organic Remains, and of British and Roman Reliques, in the Caves of Kents Hole, Anstis Cove, Chudleigh, and Berry Head*. London: Simpkin, Marshall, and Co.

Owen, R. (1846) *A History of British Fossil Mammals and Birds*. London: John Van Voorst.

Pengelly, H., Pengelly, W. (1897) *A Memoir of William Pengelly of Torquay, F. R. S., Geologist, with a Selection from His Correspondence*. London: John Murray.

Reumer, J. W. F., Rook, L., Borg, K. van der, Post, K., Mol, D., Vos, J. de (2003) Late Pleistocene survival of the saber-toothed cat *Homotherium* in Northwestern Europe. *Journal of Vertebrate Palaeontology*, 23, 260–262.

Turner, A., Antón, M. (1997) *The Big Cats and Their Fossil Relatives*. New York: Columbia University Press.

White, M. J. (2017) *William Boyd Dawkins & the Victorian Science of Cave Hunting.* Barnsley: Pen and Sword Books.

Chapter 4: Cave Lion

Barnett, R., Zepeda Mendoza, M. L., Rodrigues Soares, A. E., Ho, S. Y. W., Zazula, G., Yamaguchi, N., Shapiro, B., Kirillova, I. V., Larson, G., Gilbert, M. T. P. (2016) Mitogenomics of the extinct cave lion, *Panthera spelaea* (Goldfuss, 1810), resolve its position within the *Panthera* cats. *Open Quaternary*, 2, 1–11.

Bocherens, H., Drucker, D. G., Bonjean, D., Bridault, A., Conard, N. J., Cupillard, C., Germonpre, M., Höneisen, M., Münzel, S. C., Napierala, H., Patou-Mathis, M., Stephan, E., Uerpmann, H.-P., Ziegler, R. (2011) Isotopic evidence for dietary ecology of cave lion (*Panthera spelaea*) in North-Western Europe: Prey choice, competition and implications for extinction. *Quaternary International*, 245, 249–261.

Clottes, J. (2003) *Return to Chuavet Cave. Excavating the Birthplace of Art: The First Full Report.* London: Thames & Hudson.

Cueto, M., Camaros, E., Castanos, P., Ontanon, R., Arias, P. (2016) Under the skin of a lion: Unique evidence of Upper Paleolithic exploitation and use of cave lion (*Panthera spelaea*) from the Lower Gallery of La Garma (Spain). *PLOS One*, 11, e0163591.

Kind, C., Ebinger-Rist, N., Wolf, S., Beutelspacher, T., Wehrberger, K. (2014) The smile of the Lion Man. Recent excavations in Stadel Cave (Baden-Württemberg, south-western Germany) and the restoration of the famous Upper Palaeolithic figurine. *Quartär*, 61, 129–145.

Stuart, A. J., Lister, A. M. (2010) Extinction chronology of the cave lion *Panthera spelaea. Quaternary Science Reviews*, 30, 2329–2340.

Chapter 5: Woollies

Lister, A. M., Bahn, P. (1994/2000) *Mammoths: Giants of the Ice Age.* London: Marshall Editions.

Maschenko, E. N., Potapova, O. R., Vershinina, A., Shapiro, B., Streletskaya, I. D., Vasiliev, A. A., Oblogov, G. E., Kharlamova, A. S., Potapov, E., Plicht, J. van der, Tikhonov, A. N., Serdyuk, N. V., Tarasenko, K. K. (2017) The Zhenya mammoth (*Mammuthus primigenius* (Blum.)): Taphonomy, geology, age, morphology and ancient DNA of a 48,000 year old frozen mummy from western Taimyr, Russia. *Quaternary International*, 445, 104–134.

McKay, J. J. (2017) *Discovering the Mammoth*. New York: Pegasus Books.

Nikolskiy, P., Pitulko, V.V. (2013) Evidence from the Yana Palaeolithic site, Arctic Siberia, yields clues to the riddle of mammoth hunting. *Journal of Archaeological Science*, 40, 4189–4197.

Pecnerova, P., Palkopoulou, E., Wheat, C. W., Skoglund, P., Vartanyan, S., Tikhonov, A., Nikolskiy, P., Plicht, J. van der, Diez-Del-Molino, D., Dalen, L. (2017) Mitogenome evolution in the last surviving woolly mammoth population reveals neutral and functional consequences of small population size. *Evolution Letters*, 1, 292–303.

Tolmachoff, I. (1929) The carcasses of the mammoth and rhinoceros found in the frozen ground of Siberia. *Transactions of the American Philosophical Society*, 23, 1–74.

Chapter 6: Irish Elk

Gould, S. J. (1974) The origin and function of 'bizarre' structures: Antler size and skull size in the 'Irish elk', *Megaloceros giganteus*. *Evolution*, 28, 191–220.

Hallam, J. S., Edwards, B. J. N., Barnes, B., Stuart, A. J. (1973) The remains of a Late Glacial elk with associated barbed points from High Furlong, near Blackpool, Lancashire. *Proceedings of the Prehistoric Society*, 39, 100–128.

Immel, A., Drucker, D. G., Bonazzi, M., Jahnke, T. K., Munzel, S. C., Schuenemann, V. J., Herbig, A., Kind, C.-J. K., Krause, J. (2015) Mitochondrial genomes of giant deers suggest their late survival in Central Europe. *Scientific Reports*, 5, 10853.

Lister, A. M., Edwards, C. J., Nock, D. A. W., Bunce, M., Pijlen, I. A. van, Bradley, D. G., Thomas, M. G., Barnes, I. (2005) The phylogenetic position of the 'giant deer' *Megaloceros giganteus*. *Nature*, 438, 850–853.

Noe-Nygaard, N. (1975) Two shoulder blades with healed lesions from Star Carr. *Proceedings of the Prehistoric Society*, 41, 10–16.

Stuart, A. J., Kosintev, P., Higham, T. F. G., Lister, A. M. (2004) Pleistocene to Holocene extinction dynamics in giant deer and woolly mammoth. *Nature*, 431, 684–689.

Chapter 7: Bovids

Driessen, C., Lorimer, J. (2016) Back-breeding the aurochs: The Heck brothers, National Socialism and imagined geographies for nonhuman Lebensraum. In eds P. Giaccaria & C. Minca *Hitler's Geographies*. Chicago: University of Chicago Press.

Massilani, D., Guimaraes, S., Brugal, J. P., Bennett, E. A., Tokarska, M.,
    Arbogast, R. M., Baryshnikov, G., Boeskorov, G., Castel, J. C.,
    Davydov, S., Madelaine, S., Putelat, O., Spasskaya, N. N.,
    Uerpmann, H. P., Grange, T., Geigl, E. M. (2016) Past climate
    changes, population dynamics and the origin of Bison in Europe.
    BMC Biology, 14:93, 1–17.
Nielsen, S., Andersen, J. H., Baker, J. A., Christensen, C., Glastrup,
    J., Grootes, P. M., Huls, M., Jouttijarvi, A., Larsen, E. B., Madsen,
    H. B., Muller, K., Nadeau, M., Rohrs, S., Stege, H., Stos, Z. A.,
    Waight, T. E. (2005) The Gundestrup cauldron. *Acta
    Archaeologica*, 76, 1–58.
Soubrier, J., Gower, G., Chen, K., Richards, S. M., Llamas, B., Mitchell,
    K. J., Ho, S. Y. W., Kosintsev, P., Lee, M. S. Y., Baryshnikov, G.,
    Bollongino, R., Bover, P., Burger, J., Chivall, D., Cregut-Bonnoure,
    E., Decker, J. E., Doronichev, V. B., Douka, K., Fordham, D. A.,
    Fontana, F., Fritz, C., Glimmerveen, J., Golovanova, L. V., Groves, C.,
    Guerreschi, A., Haak, W., Higham, T., Hofman-Kaminska, E.,
    Immel, A., Julien, M. A., Krause, J., Krotova, O., Langbein, F., Larson,
    G., Rohrlach, A., Scheu, A., Schnabel, R. D., Taylor, J. F., Tokarska,
    M., Tosello, G., Plicht, J., van der, Loenen, A., van, Vigne, J. D.,
    Wooley, O., Orlando, L., Kowalczyk, R., Shapiro, B., Cooper, A.
    (2016) Early cave art and ancient DNA record the origin of
    European bison. *Nature Communications*, 7, 13158.
Vuure, C. van (2005) *Retracing the Aurochs: History, Morphology and
    Ecology of an Extinct Wild Ox*. Bulgaria: Pensoft.

Chapter 8: Bears

Barlow, A., Cahill, J. A., Hartmann, S., Theunert, C., Xenikoudakis, G.,
    Fortes, G. G., Paijmans, J. L. A., Rabeder, G., Frischauf, C., Grandal-
    d'Anglade, A., Garcia-Vazquez, A., Murtskhvaladze, M., Saarma, U.,
    Anijalg, P., Skrbinsek, T., Bertorelle, G., Gasparian, B., Bar-Oz, G.,
    Pinhasi, R., Slatkin, M., Dalen, L., Shapiro, B., Hofreiter, M. (2018)
    Partial genomic survival of cave bears in living brown bears. *Nature
    Ecology and Evolution*, 2, 1563–1570.
Chaix, L., Bridault, A., Picavet, R. (1997) A tamed brown bear (*Ursus
    arctos* L.) of the Late Mesolithic from La Grande-Rivoire (Isere,
    France)? *Journal of Archaeological Science*, 24, 1067–1074.
Edwards, C. J., Suchard, M. A., Lemey, P., Welch, J. J., Barnes, I., Fulton,
    T. L., Barnett, R., O'Connell, T. C., Coxon, P., Monaghan, N.,
    Valdiosera, C. E., Lorenzen, E. D., Willerslev, E., Baryshnikov, G. F.,

Rambaut, A., Thomas, M. G., Bradley, D. G., Shapiro, B. (2011) Multiple hybridizations between ancient brown and polar bears and an Irish origin for the modern polar bear matriline. *Current Biology*, 21, 1251–1258.

Fortes, G. G., Grandal-d'Anglade, A., Kolbe, B., Fernandes, D., Meleg, I. N., Garcia-Vazquez, A., Pinto-Llona, A. C., Constantin, S., Torres, T. J. de, Ortiz, J. E., Frischauf, C., Rabeder, G., Hofreiter, M., Barlow, A. (2016) Ancient DNA reveals differences in behaviour and sociality between brown bears and extinct cave bears. *Molecular Ecology*, 25, 4907–4918.

Germonpre, M., Hamalainen, R. (2007) Fossil bear bones in the Belgian Upper Paleolithic: The possibility of a proto bear-ceremonialism. *Arctic Anthropology*, 44, 1–30.

Kurtén, B. (1976) *The Cave Bear Story: Life and Death of a Vanished Animal*. New York: Columbia University Press.

Pacher, M., Stuart, A. J. (2008) Extinction chronology and palaeobiology of the cave bear (*Ursus spelaeus*). *Boreas*, 38, 189–206.

Wojtal, P., Wilczynski, J., Nadachowski, A., Munzel, S. C. (2015) Gravettian hunting and exploitation of bears in Central Europe. *Quaternary International*, 359, 58–71.

Chapter 9: Northern Lynx

Blake, M., Naish, D., Larson, G., King, C. L., Nowell, G., Sakomoto, M., Barnett, R. (2014) Multidisciplinary investigation of a 'British big cat': A lynx killed in southern England *c.* 1903. *Historical Biology*, 26, 441–448.

Hetherington, D. A., Geslin, L. (2018) *The lynx and us*. SCOTLAND: The Big Picture.

Hetherington, D. A., Lord, T. C., Jacobi, R. M. (2005) New evidence for the occurence of Eurasian lynx (*Lynx lynx*) in medieval Britain. *Journal of Quaternary Science*, 21, 3–8.

Molloy, D. (2011) *Wildlife at work: The economic impact of white-tailed eagles on the Isle of Mull*. Sandy, UK: RSPB.

Nowell, K., Jackson, P. (1996) *Wild Cats: Status Survey and Conservation Action Plan*. Gland, Switzerland and Cambridge, UK: IUCN/SSC Cat Specialist Group.

Raye, L. (2016) The forgotten beasts in Medieval Britain: a study of extinct fauna in medieval sources. PhD thesis, Cardiff University, 397.

Chapter 10: Grey Wolf

Aybes, C., Yalden, D. W. (1995) Place-name evidence for the former distribution and status of wolves and beavers in Britain. *Mammal Review*, 25, 201–227.

Crumley, J. (2010) *The Last Wolf*. Edinburgh: Birlinn Limited.

Darwin, C. R. (1839) *The Voyage of the Beagle*. New York: Penguin.

Freedman, A. H., Wayne, R. K. (2017) Deciphering the origin of dogs: From fossils to genomes. *Annual Review of Animal Biosciences*, 5, 281–307.

Harting, J. E. (1880) *British Animals Extinct within Historic Times: With Some Account of British Wild White Cattle*. Edinburgh: Ballantyne Press.

Marvin, G. (2012) *Wolf*. London: Reaktion Books.

Scott, P. A., Bentley, C. V., Warren, J. J. (1985) Aggressive behavior by wolves toward humans. *Journal of Mammalogy*, 66, 807–809.

Wilson, P. I. (1999) Wolves, Politics, and the Nez Perce: Wolf recovery in Central Idaho and the Role of Native Tribes. *Natural Resources Journal*, 39, 543–564.

Chapter 11: Eurasian Beaver

Campbell, R. D., Harrington, A., Ross, A., Harrington, L. (2012) Distribution, population assessment and activities of beavers in Tayside. *Scottish Natural Heritage Commissioned Report No. 540*, Inverness.

Coles, B. 2006. *Beavers in Britain's Past*. Barnsley: Oxbow books.

Conroy, J. W. H., Kitchener, A. C. (1996) The Eurasian beaver (*Castor fiber*) in Scotland: A review of the literature and historical evidence. *Scottish Natural Heritage Review*, 49, 1–23.

Gaywood, M. J., Stringer, A., Blake, D., Hall, J., Hennessy, M., Tree, A., Genney, D., Macdonald, I., Tonhasca, A., Bean, C., McKinnell, J., Cohen, S., Raynor, R., Watkinson, P., Bale, D., Taylor, K., Scott, J., Blyth, S. (2015) *Beavers in Scotland: A report to the Scottish Government*. Inverness: Scottish Natural Heritage.

Jones, S., Campbell-Palmer, R. (2014) *The Scottish Beaver Trial: The story of Britain's first licensed release into the wild*. Edinburgh, UK: Scottish Wildlife Trust and Royal Zoological Society of Scotland.

McEwing, R., Senn, H., Campbell-Palmer, R. (2015) Genetic assessment of free-living beavers in and around the River Tay catchment, east Scotland. *Scottish Natural Heritage Commissioned Report No. 682*, Inverness.

Chapter 12: The Future

Barlow, C. (2000) *The Ghosts of Evolution: Nonsensical Fruit, Missing Partners, and Other Ecological Anachronisms*. New York: Basic Books.

Pilcher, H. 2016. *Bring Back the King: The New Science of De-extinction*. London: Bloomsbury Sigma.

Seddon, P. J., Moehrenschlager, A., Ewen, J. (2014) Reintroducing resurrected species: Selecting De-extinction candidates. *Trends in Ecology & Evolution*, 29, 140–7.

Shapiro, B. (2015a) *How to Clone a Mammoth*. Princeton, New Jersey: Princeton University Press.

Vera, F. W. M. (2009) Large-scale nature development – the Oostvaardersplassen. *British Wildlife*, 29, 28–36.

Zimov, S. A. (2005) Essays on science and society. Pleistocene Park: Return of the mammoth's ecosystem. *Science*, 308, 796–8.

Index